ETOPS

Extended Twin-Engine Operations

The Development of Twin-Engine Long-Haul Flight

ETOPS

Extended Twin-Engine Operations

The Development of Twin-Engine Long-Haul Flight

C B Holland

First published in Great Britain in 2016 by

Bannister Publications Ltd
118 Saltergate
Chesterfield
Derbyshire S40 1NG

Copyright © C. B. Holland

ISBN 978-1-909813-26-7

C. B. Holland asserts the moral right to be identified
as the author of this work

A catalogue record for this book is available from the British Library

This book is sold subject to the condition that it shall not, by way of trade or otherwise, be lent, re-sold, hired out or otherwise circulated without the copyright holder's prior consent in any form of binding or cover other than that in which it is published and without a similar condition including this condition being imposed on the subsequent purchase.

All rights reserved. No part of this book may be reproduced or transmitted in any form or by any means, electronic or mechanical including photocopying, recording or by any information storage and retrieval system, without permission from the copyright holder, in writing.

Typeset in Palatino Linotype by Escritor Design, Bournemouth

Printed and bound by SRP Ltd, Exeter

Contents

Introduction vi

Chapter 1 1
    The Early Pioneers

Chapter 2 11
    Post-World War 1 Propeller-powered Aircraft Piston Engine Developments

Chapter 3 39
    The Early Military Jet Engine Era

Chapter 4 59
    The Civil Jet Engine Powered Aircraft Pre-ETOPS Era

Chapter 5 83
    Civil Jet Engine Powered Aircraft

Chapter 6 109
    Early-ETOPS and Beyond

Conclusion 127

Appendix 1 129
    Notes 129
    Units, Conversion Factors 137

Appendix 2 139
    Tables 139

Bibliography and References 144

Introduction

Aircraft and aero-engines are sophisticated pieces of equipment, at the forefront of technology, yet it is inevitable that component or system failures will occasionally occur. Ground and flight testing is aimed at identifying and rectifying any weaknesses, so that in-flight failure rates can be maintained at an acceptable level. It is commonplace now to fly on long-haul flights in twin-engine aircraft, and passengers can be excused for looking out of the cabin window and wondering what happens if one of the engines should fail.

In order that twin-engine aircraft can fly on extended long-haul flights, they have to satisfy strict rules to ensure that the risk of a double engine failure during such a flight is extremely improbable. The Aircraft and Engine combination has to comply with the certification requirements for extended range operations, known as ETOPS, which determines the permitted diversion time to an alternate airport in the event of a single engine failure in flight.

A critical feature of the permitted diversion time is the prevailing engine in-flight shutdown rate (IFSDR) based upon the service experience of the total fleet of engines. This diversion time has increased over the years as engine reliability has improved; by 1988 a diversion time of 180 minutes, known as 180- minutes ETOPS, had been approved on several aircraft types, based upon an achieved fleet engine IFSDR of 0.02/1,000 hours (one every 50,000 hours).

In the mid-1980s Boeing were designing a wide-bodied, twin-engine, long range aircraft, to bridge the gap between its Boeing 767 and 747-400. It would be more fuel efficient and offer the airlines a reduction in operating costs relative to the three- and four-engine aircraft currently in service. For this proposal to be attractive, Boeing recognised that customers might not be prepared to wait for the Aircraft and Engine combination to demonstrate the in-service levels of reliability required to gain 180-minutes ETOPS approval needed for the long-haul routes.

It was therefore proposed, in agreement with the certification authorities, to devise a method of engine and aircraft design and development that would enable 180-minutes ETOPS approval to be granted at entry into service (EIS). Engine manufacturers Rolls-Royce, Pratt & Whitney and General Electric all had orders for different customers respectively with the Trent 800, PW4074 and GE 90 engines; after extensive development and flight testing, they all received FAA approval for 180-minutes ETOPS prior to aircraft certification.

A beneficial effect of this programme has been the progressive improvement in engine reliability. In-flight shutdown rates have halved to less than 0.01/1,000 hours since the adoption of these enhanced design, test and maintenance practices, irrespective of whether the aircraft has two, three or four engines, resulting in an overall improvement in safety. Dispatch reliability has also improved as a consequence.

This book describes the evolution of commercial aircraft and engines, to a degree that now allows us to fly safely over long distances. ETOPS-approved twin-engine aircraft can today operate with diversion times up to 370 minutes, in the knowledge that the risk of a double engine failure is extremely improbable.

The first four chapters describe the evolution of aircraft and aero-engines from the first commercial passenger flight in 1914 to the latest twin-engine, wide-body aircraft in use today. The evolution of passenger carrying aircraft has been very dependent upon the gradual advance of aero-engine design, from the pre-war era of the piston engine to post-war jet engines. Jet engines were then developed from turbo-jet technology to the high bypass ratio turbofans of today, whose increase in thrust and efficiency enabled the design of the wide-body twin-engine aircraft that are so popular.

Chapter 5 deals with the subject of Extended Twin-Engine Operations, ETOPS, (sometimes jovially referred to as Engines Turn Or Passengers Swim), whose introduction was primarily aimed at regulating the operation of twin-engine aircraft so that the flight time to a suitable diversion airport, with one engine inoperative, was consistent with the risk of a second engine failure being extremely improbable.

Chapter 6 deals with the enhanced testing and operation procedures adopted to enable 180-minutes ETOPS approval to be granted at EIS on the Boeing 777. The beneficial effect this had on further improving engine and aircraft reliability, and the further development of ETOPS procedures and extensions of diversion times, is also covered.

The Appendix discusses some of the issues arising out of the narrative and includes some additional interesting facts.

Chapter 1

The Early Pioneers

The First Cross Channel Flight

The benefits of being able to fly from one land mass to another were recognized in the early days of aviation. In 1909 the British Daily Mail newspaper offered a prize of £1,000 to the first aviator to fly across the English Channel.

French aviator Louis Blériot had been developing a single-engine monoplane over the previous four years. He survived three spectacular crashes in 1907 but, undeterred, in July 1909 he went on to successfully to pilot his Blériot XI aeroplane across the English Channel from Calais to Dover, a distance of 22 miles (36.6km). The journey took 36.5 minutes, flying at approximately 45 mph (72 km/h) at an altitude of about 250 ft (76 m), and made Blériot a celebrity.

Fig 1: Blériot XI Monoplane

Blériot had tried several different engines but none had sufficient reliability for the cross channel attempt. Consequently, he contacted Alessandro Anzani, a famous motorcycle racer whose successes were due to the engines which he made, and who had recently entered the field of aero-engine manufacture. An Anzani 3-cylinder, 25 Horsepower (HP), radial fan- engine was chosen for the cross Channel flight. The fan-engine arrangement was the result of it having originally being designed to fit in the base of a motorcycle frame.

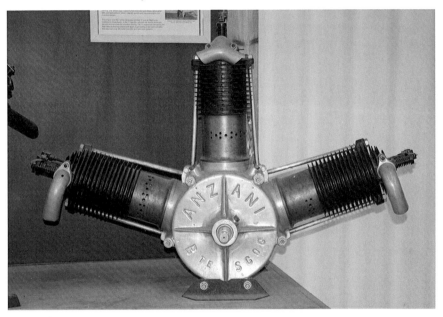

Fig 2: Anzani 3-cylinder fan-engine as used in the Blériot XI

After the successful crossing of the English Channel there was a great demand for Blériot XIs; by the end of September 1909 orders had been received for 103 aircraft. The Type XI remained in production until the outbreak of WW1, and a number of variations were produced. Blériot marketed the aircraft in four categories: trainers, sport or touring models, military aircraft, and racing or exhibition aircraft.

The Blériot XIs first entered military service in Italy and France in 1910, and a year later some of those were used by Italy in offensive action in North Africa; the first aircraft to be used in a war. The Royal Flying Corps received its first Blériot in 1912. During the early stages of WW1, eight French, six British and six Italian squadrons operated various military versions of the aircraft, mainly in observation duties but also as trainers, and in the case of single-seaters as light bombers able to carry a bomb load of up to 25 kg.

Various types of engine were fitted, including the 120-degree Y-configuration, 3-cylinder 50 HP Anzani. Later versions of the aircraft were powered by a 7-cylinder Gnome Omega rotary engine, designed and produced by the French manufacturer Société des Moteurs Gnome. (see Fig 16).

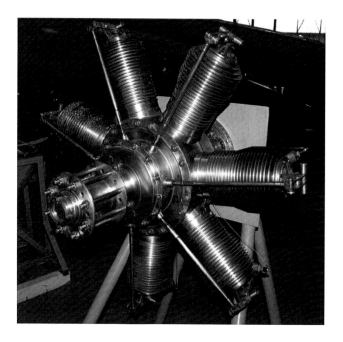

Fig 3: 7-Cylinder Le Rhône Omega Rotary Engine

The First Commercial Flight

The first known commercial airline flight took place in the USA on New Year's Day 1914. The aircraft was the Benoist XIV, a conventional biplane flying boat with equal-span unstaggered wings with small pontoons at their tips. The engine was mounted on a pedestal aft of the cockpit and drove a two-blade pusher propeller. It flew between the Florida cities of St Petersburg and Tampa, a distance of 22 miles. The pilot was a well-known aircraft pioneer Tony Jannus, the single passenger for the inaugural flight was Abram C Pheil a former mayor of St Petersburg. Mr Pheil paid $400 for the pleasure, having been successful in an auction for the only passenger seat; the official price for fare paying passengers thereafter was $5. The flight took 23 minutes.

Fig 4: Benoist XIV

The aircraft was powered by a Roberts 6-cylinder 2-stroke aero-engine which had been developed from an engine designed for boat propulsion. The engine developed 75 HP and propelled the aircraft to a maximum speed of 64 mph and a range of 125 miles.

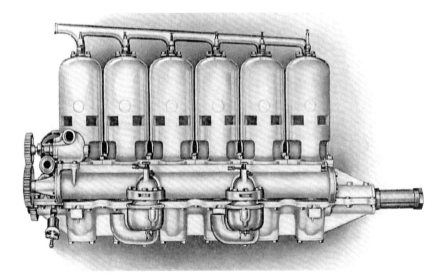

Fig 5: Roberts 6X Aero-Engine

The First Non-Stop Transatlantic Flight

The first non-stop transatlantic flight was achieved in 1919 by RAF pilots Alcock and Brown in a Vickers Vimy biplane powered by two Rolls-Royce Eagle V111-12-cylinder piston engines, the flight from Newfoundland to Galway in Ireland took three minutes under 16 hours.

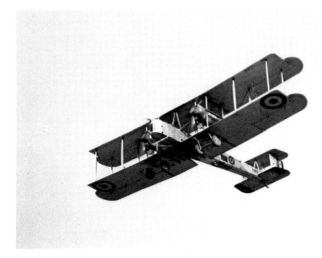

Fig 6: Vickers Vimy

The Vimy was designed as a WW1 War bomber and was named after the Battle of Vimy Ridge, a military engagement fought primarily as part of the Battle of Arras, in the Nord-Pas-de-Calais region of France. The original requirement was for a twin-engine biplane capable of attacking targets in Germany at night. The first prototype was flown in November 1917 and was powered by two 200 HP Hispano Suiza engines. Several different engine types were used during the flight development phase; the RAF selected the 360 HP Rolls-Royce Eagle for its operational aircraft. Only three aircraft had been delivered by October 1918 and the war ended before the Vickers Vimy could be used for the purpose for which it was designed. However, the aircraft formed the main heavy bomber force of the RAF for much of the 1920s and served as a front line bomber in the Middle East and in the United Kingdom from 1919 until 1925.

In April 1913 the London Daily Mail offered a prize of £10,000 to the first aviator to cross the Atlantic in an aeroplane flying from any point in the USA, Canada or Newfoundland, to any point in Great Britain or Ireland, in 72 continuous hours. The competition was suspended on the outbreak of war in 1914 but reopened after Armistice was declared in 1918.

John Alcock was a military pilot during WW1, taken prisoner in Turkey after the engines on his Handley Page bomber failed over the Gulf of Xeros (see Fig 9). After the war, he took up the challenge of attempting the first flight directly across the Atlantic. He approached Vickers who had considered entering their Vickers Vimy Mark IV twin-engine bomber in the competition, but had not yet found a pilot. Alcock's enthusiasm impressed the Vickers' team and he was appointed as their pilot. Work began on converting the Vimy for the long flight, replacing the bomb carriers with extra petrol tanks.

Arthur Whitten Brown was a navigator who had been shot down over Germany in the war. He was also employed by Vickers after the war as a navigator. The two of them were selected to be the flight crew for the transatlantic challenge.

The Vickers Vimy Mark IV was powered by two Eagle V111 engines, loaned by Rolls-Royce to Vickers specifically for the transatlantic flight. The engines were specially prepared for the flight and modifications were incorporated to address a few known problems. Special fuel was used, refined by Shell; filtered, distilled water was used to minimise the risk of clogging in the radiator.

The normal range of the aircraft was 900 miles; this was extended by the replacement of the bomb bays with fuel tanks, increasing the fuel capacity to 856 gallons. The normal operating maximum altitude with a full complement of bombs was 7,000 ft and this was increased to 11,000 ft, due to the lower aircraft weight.

The aircraft took off from Lester's Field, St Johns, Newfoundland on 14 June 1919, and after a flight of 15 hrs 57 min landed in a peat bog in Galway, Ireland.

The flight was not without incident. Failure of the exhaust manifold on the starboard engine early on in the flight resulted in deafening noise making communication extremely difficult. The electrical heating of the pilots' suits ceased to work after five hours due to the battery running low. Icing was a continual problem, and Brown bravely scrambled out on to the wing several times to chip away at accumulations. Fog and low cloud made navigation difficult; this was exacerbated by failure of a generator which prevented wireless direction finding. Extreme turbulence resulted in an out of control descent of thousands of feet to very close to sea level.

Notwithstanding these difficulties the flight satisfied the conditions of the Daily Mail challenge, and Alcock and Brown received their share of the £10,000 prize. Both were knighted for their achievement. Regrettably

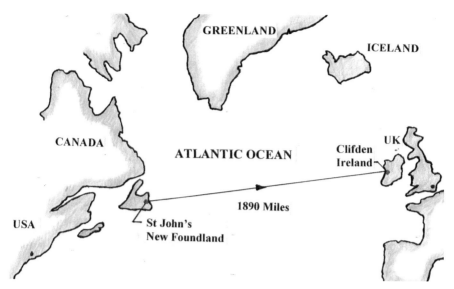

Fig 7: Transatlantic Flight Route

Sir John Alcock was killed in December 1919 when the Vickers Viking Amphibian he was piloting to the Paris Exhibition crashed whilst attempting to land in foggy weather.

Later that year a modified Vickers Vimy, also powered by two Rolls-Royce Eagle V111 engines, became the first aircraft to fly from England to Australia. The length of the total journey was 11,294 miles and took from 12 November to 10 December and included 21 stops en route, the journey was mainly overland with the exception of crossing the Mediterranean and the Timor Sea. A commercial version of the Vimy was made which had a larger diameter fuselage made of spruce, and had seating for ten passengers.

Development of The Rolls-Royce Eagle V111 engine

Prior to 1914 Sir Henry Royce, with his colleague Charles Rolls, had set up a factory in Derby, England, for the purpose of manufacturing quality motor vehicles. Their Rolls-Royce Silver Ghost SP40/50 was their most prestigious vehicle at the beginning of WW1, this car was powered by the 6-cylinder 40/50 engine designed by Henry Royce, and manufactured in Derby.

At the beginning of WW1, Rolls-Royce agreed, after much wrangling, to build a batch of 220 Renault 80 HP aero-engines under licence for the War Office, this prompted Henry Royce to consider possible Rolls-Royce designed aero-engine options. The two potential customers were the

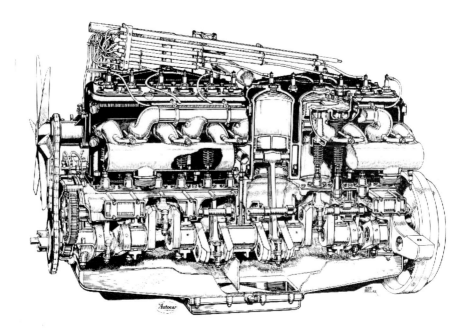

Fig 8: Rolls-Royce 40/50 Motor Car Engine

Royal Naval Air Service (RNAS) and the Royal Airforce (RAF). The RNAS specified a requirement for a 200 HP engine to suit a twin-engine patrol bomber under consideration. Henry Royce was an intuitive designer who believed in basing his designs on sound previous experience; he was quite prepared to incorporate best practice from competitive products. He quickly decided to use his successful 40/50 motor-engine as the basis of his design.

The power requirement would be achieved by arranging two rows of six cylinders side by side in a 60-degree V formation. The pressure ratio was increased from 3.5:1 to 4.5:1, the engine capacity (swept volume) was increased from 453 cubic inches (7.43 litres) to 1,240.5 cubic inches (20.33 litres), the piston stroke was increased from 4.75 inches to 6.5 inches and the crankshaft speed was increased from approximately 1,250 rpm to 1,800 rpm. The engine would be water cooled which involved encasing the cylinders in fabricated water jackets. Royce considered the overhead valve arrangement of the Mercedes DF 80 aero- engine to be superior to the side valve arrangement of 40/50 engine, and had no hesitation in incorporating the technology, but improved by his own modifications; the crankshaft 7-bearing arrangement of the 40/50 car engine was retained. Significant weight reduction had to be incorporated in the design to meet the contract guarantee dry weight of 736 lb (the 6-cylinder car engine weighed just under 900lb).

The installation drawing to the agreed configuration was issued at the beginning of September 1914. The first engine run was a mere six months later at the end of February 1915, four days later 225 HP was demonstrated at 1,600 rpm. Royce had deliberately designed the engine with some growth potential; he had already demonstrated the ability of the 6-cylinder 40/50 car engine to achieve 80 HP at 2,250 rpm. In addition to demonstrating the achievement of contract requirements, it was his philosophy to subject his designs to over-stress conditions so as to highlight any weak points, sophisticated stress analyses were not available at the time and this approach enabled him to rectify any shortcomings in the design. At the end of March, the engine ran successfully for six hours at 2,000 rpm, exposing some problems that were quickly rectified. The first two flight-worthy engines were delivered to RNAS in October 1915. The first flight of the twin-engine Handley Page 0/100 Patrol Bomber powered by two RR Eagle V12 engines took place on 17 December 1915, this was the first ever flight of a Rolls-Royce aero-engine.

Fig 9: Handley Page 0/100 Patrol Bomber

It is uncertain which Mark of Eagle engine went into the first Handley Page 0/100 aircraft, however it is likely to have had a HP of at least 250, significantly above the original contract requirement.

Engine design and development continued, with a rating of 360 HP being achieved in the Eagle V12 -V111 in 1917, it was this engine that was chosen to power the Vickers Vimy Bomber for the transatlantic flight. The Eagle had a reputation for reliability, undoubtedly due to Henry Royce's philosophy of testing to failure, regrettably no reliability statistics have emerged by which to compare the Eagle with other competing engines, however it appears to have been generally accepted that it was superior to its peers.

In spite of this there was an occasion when it featured in a double-engine failure; on 30 September 1917, John Alcock was piloting a Handley Page 0/100 bomber powered by Rolls-Royce Eagle engines, when he was forced to turn back to base after an engine failed near Gallipoli. After flying on the single engine for more than 60 miles, that engine also failed and the aircraft ditched in the sea near Suvla Bay. Alcock and his crew of two were unable to attract the attention of nearby British destroyers, and when the plane finally began to sink they swam for an hour to reach the enemy-held shore; all three were taken prisoner next day by the Turkish forces. Alcock remained a prisoner of war until the Armistice and retired from the Royal Air Force in March 1919. The engine failures were believed to be due to a common cause, i.e. failure of an oil supply pipe. There is no evidence that these pipes were part of the engine equipment, it seems more likely that they were aircraft supply from an auxiliary oil tank that was mounted in the aircraft; if so, the problem would not have been exposed by engine testing at Rolls-Royce. Oil consumption was in the region of one gallon per hour. In spite of this experience, Alcock was not deterred from participating in his transatlantic crossing in the Eagle powered Vickers Vimy in 1919.

It is interesting that these three pioneering flights were respectively powered by developments of a motorcycle, motor boat and motor car engine.

Chapter 2

Post-World War 1 Propeller-powered Aircraft Piston Engine Developments

In-line Engines

Rolls-Royce continued to develop the V12 piston engine after the end of WW1, primarily for military applications. The Kestrel, a 21 litre capacity supercharged engine, was introduced in 1927. It was a state-of-the-art engine, having a cast aluminium cylinder block and a pressurised cooling system. Prior water-cooled engines had a maximum operating altitude limited by the boiling point of the cooling water, which decreases with altitude as the pressure reduces.

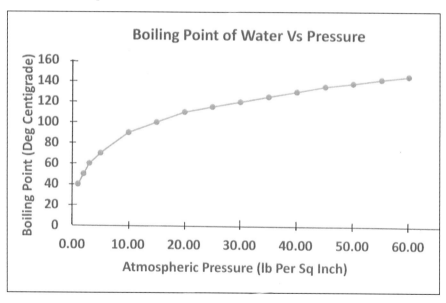

Pressurising the cooling system to 3Bar (44.1psi) enabled it to operate up to 135 degrees centigrade before any boiling occurred, effectively removing the altitude restriction.

The prototype Kestrel first ran in 1926 and powered an aircraft the following year. The engine's power output was initially 450 HP (340kW), and this variant saw widespread use in the Hawker Hart family that was the mainstay of British air power during the early 1930s. Power was eventually boosted to 720 HP without any change to the basic configuration by the use of higher octane fuel. The higher power, lower weight and reliability of the RR Kestrel made it the first choice power plant for a range of military aircraft; it was even chosen to power the prototypes of the German Messerschmitt BF 109 fighter and the Junkers JU 87 Stuka dive bomber. The Rolls-Royce Peregrine, a larger variant of the Kestrel, was introduced in 1938 to increase the power rating to 885 HP.

Fig 10: Rolls-Royce Kestrel

In the early 1930s, realising there was a need for a larger engine than the Kestrel, Rolls-Royce started planning its future aero-engine development programme. Work was started on a new 1,100 HP (820 kW) engine. As the company received no government funding for work on the project the design was designated the PV-12, standing for *Private Venture 12-cylinder*. The PV-12 was first run on 15 October 1933, and first flew in a Hawker Hart biplane on 21 February 1935. In that year the British Air Ministry issued a specification for a new fighter aircraft with a minimum airspeed of 310 mph (500 km/h). Two designs were produced based around the PV-12 engine: the Supermarine Spitfire and the Hawker Hurricane. Production contracts were placed in 1936 and the engine was renamed the Merlin.

These engine developments were primarily geared towards military aircraft, which is not surprising considering the looming prospect of war.

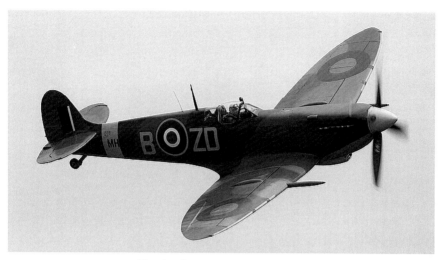

Fig 11: Supermarine Spitfire

In parallel, significant developments were taking place with radial engines in both military and civil applications.

The Rolls-Royce Griffon, a larger, 27 litre capacity version of the Merlin, producing up to 2,200 HP, was fitted to later versions of the Spitfire.

The Merlin was successfully fitted in many other aircraft types, not least the Avro Lancaster and De Havilland Mosquito. After the war the civil aviation industry restarted with Merlin-powered Lancastrian and York transports and the Canadian-built DC-4M Argonaut, many of which supported the Berlin Airlift in 1948.

Fig 12: Canadair DC4 Argonaut with RR Merlin Engines

By the end of production in 1951 a total of 168,040 Merlin engines had been produced, including over 55,000 built under licence as the V-1650 by Packard in the US. The Griffon was the last in the line of V-12 aero-engines to be produced by Rolls-Royce, with production ceasing in 1955.

There appears to be no reliable data on failure rates for in-line piston engines. Civil aircraft applications were in small quantities, and military records are not relevant due to the type of operation.

Radial Engine Developments: Radial Piston Engines

They say there is more than one way of skinning a cat, and in the case of piston engines there are two choices, in-line or radial cylinders. In-line engines have their cylinders either straight in line, or arranged in a V formation as in the Rolls-Royce Eagle and its derivatives described above. In a radial engine the cylinders radiate outward from a central crankcase like the spokes of a wheel. Unlike the in-line engine the axes of the cylinders are co-planar, making it difficult to attach the piston connecting rods directly to the crankshaft unless mechanically complex forked connecting rods are used. To avoid this complexity, the pistons are ingeniously connected to the crankshaft with a master and articulating rod assembly. One piston has a master rod with a direct attachment to the crankshaft, the remaining pistons pin their connecting rod ('con-rod') attachments to the inner end of the master rod. Four stroke radial engines have an odd number of cylinders per row, this enables a firing order where every other piston fires providing a smooth sequence of operation. For a 5-cylinder engine the firing order is 1, 3, 5, 2, 4 and back to cylinder 1. This firing order leaves a one-cylinder gap between the piston on its combustion stroke and the piston on its compression stroke. The active stroke directly helps compress the next cylinder to fire, making the motion more uniform, this smoothness of operation is not achievable with an equal number of cylinders. The following diagram of a 9-cylinder radial engine shows the master con-rod at top dead centre (TDC), with the other eight con-rods attached to it.

The inner end of the master rod rotates eccentrically around the crankshaft centreline and is offset by a counter-balance weight. An excellent working animation of Fig 13 can be viewed on the *Mekanizmalar* website.

Each configuration had its advantages and disadvantages. In-line engines had a much smaller cross-section, which simplified streamlining.

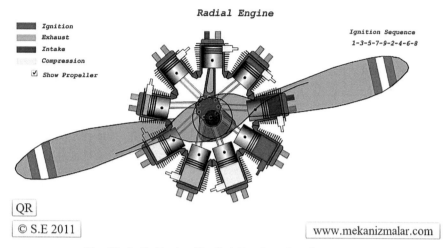

Fig 13: 9-Cylinder Radial Engine Configuration

However, the in-line cylinder layout could not be adequately cooled by airflow alone, so a liquid cooling system (water) was required. This was complex to apply and was vulnerable to leakage due to battle damage. Radial engines had a large frontal area which increased drag relative to the in-line engine. This became more severe at higher speeds, as drag is proportional to the square of the speed. However, with the application of cooling fins, the co-planar cylinders could be more easily cooled by air from the upstream propeller. The trade-off was greater drag for the radial engine, versus greater weight and vulnerability to damage for the in-line engine. Aircraft employing radial engines had a blunt nose due to the

Fig 14: Mustang P-51 with in-line engine

Fig 15: Corsair with radial engine

large cross-section of the engine, while aircraft employing in-line engines had a pointed, streamlined nose to take maximum advantage of the reduced engine cross-section.

Ultimately the radial engine had greater potential for higher power with the addition of a second row of cylinders, which became crucially important in the development of engines to power large airliners.

Later versions of the Blériot XI aircraft were powered by a 7-cylinder Gnome Omega rotary engine, designed and produced by the French manufacturer Société des Moteurs Gnome. The rotary engine was similar to the radial engine, the difference being that in the rotary engine the cylinders and crankcase rotated around a fixed crankshaft mounted to the airframe, whereas in the radial engine the crankshaft rotates within the fixed crankcase. The engines look very similar externally. Early radial engines tended to overheat, particularly at low power where air velocity was low. The rotation of the cylinder block helped to alleviate this problem but had the disadvantage that rotation of such a large mass at high rpm resulted in high gyroscopic forces. During level flight the effect was not especially apparent; however, when turning, the effect of gyroscopic precession became noticeable (the motion of a spinning body in which it wobbles so that the axis of rotation sweeps out a cone). Due to the direction of the engine's rotation, left turns required effort and

Fig 16: Installed Gnome Omega 7-cylinder Rotary Engine

happened relatively slowly, combined with a tendency to nose up, while right turns were almost instantaneous, with a tendency for the nose to drop.

These rotary engines proved very effective during WW1 in many applications. However, the gyroscopic effect limited the development of the rotary engine, as increased power required higher rpm and increased mass, both of which exacerbated the problem. Also the power loss associated with the windage generated by the rotating mass seriously limited the ability to develop more horsepower. The introduction of thin cooling fins on the exterior of cylinder walls enabled pure radial engines to largely overcome the overheating problems and by the end of WW1 they had virtually superseded the rotary engine.

The Gnome Omega was introduced in the spring of 1909 with a capacity of 488 cubic inches (8 litres), delivering 50 HP (37 kW) at 1,200 rpm with a dry weight of 165lb (75 kg). The Gnome powered a range of different aircraft including Henry Farman's Farman 111, which took world records for distance and endurance, as well as powering the first aircraft to exceed 100 km/h. In 1910 it also powered the Fabre Hydravion, the first seaplane ever to fly. More than 1,700 of these engines were built in France, along with models built under licence in Germany, Sweden, Britain, the United States and Russia.

In 1911 the basic Gnome Omega was followed by the Gnome Lambda, a larger 7-cylinder version of 80 HP (60 kW). This was followed in 1914 by the 9-cylinder 100 HP (75 kW) Gnome Delta (also called the Gnome Monosoupape, or single valve, as it used that type of engine design for the first time). Gnome also tried a 14-cylinder two-row version, the Double Lambda of 160 HP (120 kW), the first known radial-configuration engine to ever use a twin-row design, but it saw little use. It was copied by Oberursel in Germany as the U.III, and used in a few early Fokker fighter designs, but without success. To deliver more power and to compete with high power in-line engines developed late in the war, a completely new 9-cylinder 160 HP Monosoupape design was delivered in 1918 as the Type-N. This design was used on the Nieuport 28, a French biplane, of which 297 were purchased by the US at the end of WW1.

Another French engineer, Louis Verdet, designed his own small rotary engine in 1910. It saw little use and in 1912 he delivered a larger 7-cylinder design, the 7C, which developed 70 HP and proved much more popular. He formed Société des Moteurs Le Rhône later that year. He soon followed the 7C with a 9-cylinder design, the Le Rhône 9C, delivering 80 HP (60 kW). These engines were direct competitors for the Gnome Omega and Lampda. Like Gnome, the Le Rhône designs were widely licensed; the 110 HP Le Rhône 9J was produced in Germany by Oberursel, whose Le Rhône engine copies received a "Ur" prefix. In the US the Union Switch & Signal Company of Swissvale, Pa. was reported to have produced some 10,000 units, as well as Austria, Britain and Sweden.

After several years of fierce competition, Gnome and Le Rhône finally decided to merge. Negotiations started in 1914 and on 12 January 1915 Gnome bought out Le Rhône to form Société des Moteurs Gnome et Rhône. Developments of the 9C continued to be their primary product, by the end of the war improving in power to about 110 HP (80 kW) in the Le Rhône 9J. The 9-series was the primary engine for most of the early WW1 designs, both in French and British service as well as in Germany. Between 1914 and 1918 they produced 25,000 of their 9-cylinder Delta and Le Rhône 110 HP (81 kW) rotary designs, while another 75,000 were produced by various licensees, powering the majority of aircraft in the first half of the war on both sides of the conflict. At the end of the war the company diversified producing a number of products, but by 1920 their rotary engines were no longer competitive, and they had no new designs in the pipeline.

In 1921 they took out a licence to manufacture the Bristol Jupiter, a

9-cylinder radial engine produced by the Bristol Aerospace Company, England. The Jupiter was designed during WW1 by R Fedden of Cosmos Engineering. The company became insolvent in 1920 due to the rapid downscaling of military spending after the war, and was subsequently purchased by the Bristol Aeroplane Company. The Jupiter engine matured into one of the most reliable on the market, and was the first to be fitted to passenger carrying airliners. The initial engines produced 400 HP, and later engines 580 HP. The engine was further developed into the Mercury and Pegasus versions. The Jupiter was used in a large variety of aircraft. Almost 21,000 engines were produced, with a number also being built in Europe under licence. The Bristol Mercury was used to power both civil and military aircraft of the 1930s and 1940s. Developed from the earlier Jupiter engine, later variants could produce 800 HP (600 kW) from its capacity of 1,500 cubic inches (25L) by use of a geared supercharger. The Pegasus was a development of the Mercury, and supercharged variants eventually produced 1,000 HP (750 kW) from a capacity of 1,750 cubic inches (28 L).

In 1922, Paul-Louis Weiller, a WW1 ace, took over the Société des Moteurs Gnome et Rhône company and decided to rekindle its interest in aircraft engines. With the experience gained from manufacturing the Bristol Jupiter they introduced their own 9-cylinder design. In 1926 they took out a licence for the smaller 5-cylinder Bristol Titan, while in return Bristol licensed the Farman-style reduction gearing used by Gnome.

Gnome-Rhône was not satisfied with simply producing Bristol designs under licence, but started a major design effort based around the mechanical features of the Titan engine. The results were introduced in 1927 as the K-series, spanning the 260 HP (190 kW) Gnome-Rhône 5K Titan, the 7-cylinder 370 HP (270 kW) the Gnome Rhône Titan Major, and the 9-cylinder 550 HP (405 kW) Gnome Rhône 9K Mistral. With the introduction of the K-series, Gnome-Rhône finally ended royalty payments to the Bristol Aerospace Company and the Gnome-Rhône 5K was built in much greater numbers than the original Bristol Titan. By 1930 they had delivered 6,000 Jupiters, Mistrals and Titans, making them the largest engine company in France.

To satisfy the demand for more power, Gnome et Rhône resurrected the two-row 14-cylinder design which they continued to develop into the Mistral Major a 14-cylinder, two-row, air-cooled radial engine. It first ran in 1929 and was Gnome et Rhône's major aircraft engine prior to WW2.

It matured into a highly sought-after design that would see licensed production throughout Europe and Japan. Thousands of Mistral Major engines were produced for use on a wide variety of aircraft.

The Bristol Engine Company later built a more powerful 14-cylinder two-row radial engine, the Bristol Hercules, which was available in 1939 and powered a wide range of military and civil aircraft; later versions delivered up to 1,735 HP. The engine was unique in that it used a sleeved valve arrangement rather than conventional poppet valves. It was very successful and a total of 57,400 were built.

Fig 17: Bristol Hercules, 14-Cylinder, Twin-row Radial Engine

Although the European engines were very successful, they failed to enjoy the same impact on the civil aircraft market as the competing engines produced by the American engine manufacturers Curtiss Wright, and later Pratt & Whitney. European engines were not inferior, but were victims of the failure of the European aircraft companies to compete with the American aircraft companies Boeing, McDonnell Douglas and Lockheed, who were naturally biased towards their own aero-engine industry.

After the end of WW1, the Americans were concerned that none of their domestic aero-engine manufacturers were actively developing a viable air cooled radial engine. In 1920 the Army Air Service announced a competition for the design and construction of an air cooled radial engine for pursuit planes. In 1921 the Lawrance Aero Engine Company, founded by engine pioneer Charles Lawrance, submitted its design for the 9-cylinder, 160 HP, Lawrence L-1, which was the first engine of that type to pass the Army's 50-hour endurance test. The engine was soon uprated to 200 HP and designated the J-1.

The US Navy was very enthusiastic about air cooled radials and ordered 50 of the uprated engines for training purposes. The Navy badly needed light reliable engines for its carrier borne aircraft, but was concerned that Lawrance could not produce enough engines for its needs. In the Navy's judgment, the Wright Aeronautical Corporation, set up to manufacture Hispano Suiza engines under licence, was better suited to meet its needs. In 1923 the Navy therefore persuaded Hispano Suiza to purchase the Lawrance company and build the J-1 itself. Wright's J-1 was the first engine in its 9-cylinder, 220 HP, R-790 Whirlwind series and was quickly followed by the J-3, J-4, J-4A, J-4B, and finally the popular and successful J-5 of 1925.

In 1924, Wright Aeronautical Corporation President Frederick Brandt Rentschler fell out with his co-directors and left the company. Together with Wright's chief engineer George Meade, Rentschler developed a proposal for a high-powered air cooled aircraft engine for the US Navy. In 1925 he set up the Pratt & Whitney Aircraft Company in Hartford, Connecticut, as a partially owned subsidiary of the Pratt & Whitney Toolmaker Company. The newly formed company immediately set about producing a new engine to compete with the Wright Whirlwind.

In the meantime, in 1926, Wright launched a new design, designated the Wright Cyclone 1750 that indicated a capacity of 1,750 cubic inches (28.67 litres). The turbocharged engine delivered 500 HP at 1,700 rpm, and was used to power various military aircraft.

In 1929 the Wright Aeronautical Company and Curtiss Aeroplane and Motor Company, merged to become the Curtiss Wright Corporation, however, the engines continued to be named after the Wright Company. The new company introduced an increased capacity version of the Cyclone-9, the R-1820, which initially had a power rating of 575 HP and was eventually developed up 1,200 HP at 2,500 rpm.

The Wright 1820 cyclone was introduced in 1932 and used to power many different aircraft, the most famous of which was the iconic four-engine B17 Flying Fortress bomber, of which over 12,741 were built before production ceased in May 1945.

Fig 18: Wright Cyclone R-1820

Fig 19: Boeing B17 Flying Fortress

The First Successful Commercial Twin-Engine Aircraft

After the end of WW1 there had been a proliferation of short-haul passenger aircraft with various engine combinations. Most of these were of wooden construction and had cruising speeds of around 100 mph.

Passenger comfort was minimal and a maximum of 12 seats made it an expensive form of travel.

The development of the radial piston engines described above enabled greater power to be achieved, and facilitated the design of twin-engine airliners of sufficient size and range to be commercially viable.

Fig 20: Boeing 247

Fig 21: Pratt & Whitney R-1340 Wasp 550 HP

A major step forward came in 1933 with the introduction of the Boeing 247, an all-metal, twin-engine, retractable-gear, streamlined airliner that could carry ten passengers in air conditioned comfort. Other innovations included sound proofing, wing and tail de-icing systems, and aileron and elevator trim tabs. There was also a galley and toilet at the rear of the airplane – sheer luxury compared with what had gone before. The aircraft was powered by Two Pratt & Whitney 550 HP Wasp S1H1-G, 9-cylinder radial engines.

Cruising speed was 180 mph which was a significant increase relative to its predecessors; service ceiling was 25,400 ft and operating range was 800 miles. The higher cruising speed justified the introduction of variable pitch propellers for the first time on a commercial airliner; this enabled the optimum engine speed to be maintained during flight.

Shortly afterwards McDonald Douglas introduced their all-metal Douglas Commercial DC-1 prototype aircraft, with seating for 12 passengers and a range of 1,000 miles. The DC-1 was powered by two Wright Cyclone, 9-cylinder radial engines, each producing 710 HP. It had a cruising speed of 190 mph, yet only one aircraft was made. Following successful flight trials with the DC-1, TWA ordered a slightly larger version with 14 seats designated the DC-2. This variant had a slightly longer fuselage of increased diameter to facilitate an extra row of seats and allow taller passengers to stand in the cabin. The engines were up rated to 730 HP to compensate for the increase in weight and to provide an increase in cruising speed and range.

Fig 22: Douglas DC-2

The Douglas DC-2 production model began operations in July 1934. TWA advertised a coast-to-coast service in a 200 mph luxury airliner, which it called the *Sky Chief*. Transcontinental flights consisted of four legs from New York (Newark) to Chicago, Kansas City, Albuquerque and Los Angeles. Flights left at 4:00 p.m. and arrived at 7:00 a.m. the next day. It became the best passenger aircraft in the world and other operators soon began queuing up to place orders. The first of the non-US airline customers was Dutch Airline KLM, which began flying the type in the autumn of 1934. The DC-2 was set up for a long production run. While the European manufacturers had focused their attention on military aircraft, the Americans had been steadily improving commercial aviation; there was no European commercial equivalent to the DC-2.

Douglas continued to develop the concept, inspired by a requirement from American Airlines for an aircraft with side-by-side sleeping accommodation for 14 to 16 passengers. The 66-inch diameter DC-2 fuselage was too narrow for this purpose and a new aircraft was launched, designated the DST, with a fuselage diameter of 92 inches. This aircraft first flew in December 1935. It quickly became clear that the flying-sleeper business was not what the airlines had hoped for. Too few people could be carried to make a sleeper flight profitable, or needed to travel so far by air that they would require sleeping accommodation in flight. The aircraft could be reconfigured to a 21-seat conventional passenger carrying arrangement with minimal change; hence the iconic Douglas DC-3 aircraft was born. The DC-3 entered service with American Airlines in June 1936. It popularised air travel in the USA, eventually replacing transcontinental rail travel; eastbound transcontinental flights could cross the US in about 15 hours with three refueling stops; westbound trips against the wind took 17½ hours.

The initial DC-3 aircraft were powered by two Wright R-1820, 9-cylinder radial engines each delivering 1,100 HP (see Fig 18). Later variants were powered by Pratt & Whitney 14-cylinder Twin Wasp radial engines each delivering 1,200HP.

Fig 23: Douglas DC-3/C47 Dakota

Fig 24: Pratt & Whitney R-1830 Twin Wasp

Pratt & Whitney introduced the 14-cylinder twin-row radial engine designated the R-1830 Twin Wasp in 1932. The engine had a capacity of 1,829 cubic inches (30 litres) delivering 1,200HP at 2,700 rpm. There were many variants with horsepowers ranging between 800 and 1,350, and there were many aircraft applications, both military and civil. The most famous military application was the four-engine B-24 Liberator, a WW2 heavy bomber of which 18,400 units were produced – the greatest number of heavy bombers in history.

The Douglas DC-3 was the Wasp's most famous civil application. The aircraft had a cruising speed of 207 mph and a range of about 1,500 miles. The nominal cruising altitude was 8,000 ft, as the cabin was unpressurised. However, the aircraft had an operational ceiling of 22,700 ft, which enabled it to fly above turbulence caused by the ground, or over minor weather bumps in the sky. The improved aerodynamics and increased size and weight of the airframe relative to its predecessors resulted in a smoother ride. Passenger comfort was further improved by the incorpo-

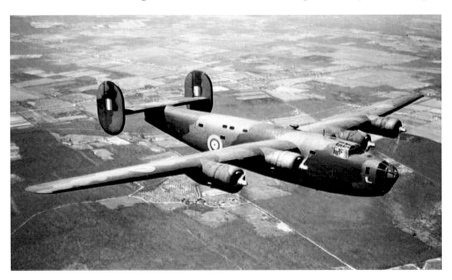

Fig 25:L B24 Liberator Heavy Bomber in RAF Livery

ration of fresh air ventilation and noise reduction, accomplished by the introduction of padded walls, carpeted floors, upholstered seats, rubber vibration-dampers and shock-mounts throughout the aircraft's body.

The introduction of the DC-3 had a dramatic effect on airline travel; the cost-per-passenger mile was much lower than its predecessors, bringing ticket prices within reach of a larger percentage of the travelling public. It was the first airliner to truly be profitable at hauling only passengers; prior airliners could only make money if they carried valuable lightweight cargo, such as mail. Between its introduction in 1935 and America's entry into WW2 at the end of 1941, there was a 600 percent growth in the number of passenger miles flown across the USA, the majority of which were flown on DC-3s. It would be years before any competitor could match the DC-3 for anywhere near the same purchase and operating costs. For a decade and a half, the DC-3 remained the pre-eminent American airliner. A military version known as the Dakota was used extensively during WW2. In all, over 16,000 DC-3s and its derivatives were made.

1936: First Regulation Restricting Route Planning for Passenger Aircraft Introduced

The rapid increase in commercial travel resulting from the success of the Boeing 247, Douglas DC-2 and DC-3, together with poor piston engine reliability, caused the Federal Aviation Authority (FAA) to have concerns regarding flight safety. As a result, in 1936 they introduced a restriction that a commercial aircraft may not fly more than 100 miles from the nearest suitable diversion airport. The rule applied to all commercial aircraft, over land or water, irrespective of the number of engines. This rule was the first step in the evolution of ETOPS.

Transatlantic Flights

It was a further 19 years before the first non-stop commercial transatlantic flight. The availability of radial piston engines of up to 1,200 HP, as used in the DC-3, enabled the development of a viable 4-engine aircraft suitable for long range-flights. The first non-stop transatlantic flight was accomplished in August 1938 by the German airline Lufthansa using a 4-engine Focke Wulf Fw 200 Condor, an all-metal aircraft which flew from Berlin Tempelhof Airport to New York's Floyd Bennett Field, a distance of 3,728 miles flown in just under 25 hours, setting a new speed and distance record for land-based passenger aircraft. The aircraft had seating for 26 passengers and was powered by four Pratt & Whitney, Hornet, 9-cylinder radial engines each producing 875 HP. The designation "Condor" was chosen because, like the condor bird, the Fw 200 had a very long wingspan to facilitate high altitude flight. The aircraft was designed to cruise at an altitude of 10,000 ft to enable it to fly long distances economically; 10,000 ft was considered the highest altitude possible without cabin pressurisation.

Fig 26: Model of Fw 200 Condor

The intervention of WW2 in 1939 curtailed the development of the aircraft for civil applications. During the war the Condor was strengthened and used for reconnaissance, and less successfully as a bomber. Later versions were powered by BMW Bramo 323-R, 9-cylinder radial engines each producing 1,200 HP. Adolf Hitler had the third prototype aircraft converted for his personal use; ironically it was destroyed at Berlin Tempelhof Airport in a bombing raid in 1944.

In the mid-1930s Pan American Airways (Pan Am) had been operating a small fleet of long range flying boats to Pacific destinations, the Sikorsky S-42 and Martin M130. Flying boats had the advantage of requiring shorter range than land based aircraft and overcame the shortage of suitable landing sites in their destination countries. Pan Am was keen to extend the range of its routes within the Pacific and also to introduce a transatlantic service. In response to this requirement Boeing designed the

314 Clipper. To meet the range requirement, Boeing used the wing design previously developed for a long range military aircraft, the XB-15.

Fig 27: Boeing 314 Clipper

The initial six Boeing 314 aircraft were powered by four Wright R-2600, Double Cyclone, 14-cylinder, 1,500 HP radial engines, and had accommodation for up to 68 passengers. Range was 3,500 miles and cruising speed was 183 mph, so that journey times were long. Accommodation was luxurious and affordable only by the rich.

Curtiss Wright began working on a more powerful version of the R-1820 in 1935; the result was the R-2600 supercharged Twin-Cyclone, with 14 cylinders arranged in two rows. The engine was produced at power ratings between 1,500 and 1,900 HP and was a direct competitor to the Pratt & Whitney R-1830 Twin-Wasp. The **R-2600** engine was selected to power the Douglas A-20 Havoc (DB7) twin-engine bomber in preference to the Pratt & Whitney R-1830 engines that had powered the earlier variants. These aircraft were made in large numbers during the war.

Early engines had a number of reliability problems that contributed to several accidents; these included high oil consumption, excessive piston wear, carburation problems and cylinder wall pitting. More than 25 percent of the engines tested failed during a 3-hour test run. The situation

Fig 28: Wright R-2600 Twin-Cyclone 14-Cylinder Engine

was so serious that it was referred to the Truman Committee for investigation. They found that there was widespread production of sub standard and defective material.

During an investigation of the Lockland Plant in Ohio, inspectors found falsification of tests and forging of inspection reports, amongst other deficiencies. The problems were duly rectified and the R-2600 went on to become a much improved engine.

On 20 May 1939, Pan Am inaugurated the first transatlantic mail service. Almost a ton of mail was carried from Port Washington to Marseilles, via the Azores and Lisbon, in 29 hours. The same aircraft opened the northern mail service to Southampton on 24 June 1939.

On 28 June 1939, Pan Am inaugurated the first regular passenger service from New York to Southampton, via Newfoundland. Under the command of Captain R.O.D. Sullivan, the 'Dixie Clipper' carried the first scheduled passengers across the north Atlantic; 22 privileged passengers had the option of paying $375 one-way or $675 return. The 'Yankee Clipper' opened the same passenger route on 8 July 1939, carrying 17 passengers at the same fare. However, the golden age of the commercial flying boats was abruptly interrupted by the outbreak of the WW2 in 1939. This curtailed Pan Am's opportunity to build on its success; the northern transatlantic route was abandoned after only three months, on 3 October 1939.

Pan Am had been successfully operating the aircraft on its Pacific routes and placed an order for another six aircraft of improved specification, with increased range and extra seating capacity. The aircraft was designated the Boeing 314A and had a carrying capacity of 77 passengers, increased engine power to 1,600HP, and increased fuel capacity of nearly 1,000 gallons (4,500 litres). Due to the war, only half the order went to Pan Am, the remaining three models were bought by the British government and allotted to BOAC for use as transport aircraft. During the war years the Pan Am aircraft were also requisitioned for military use.

Interestingly, in 1945 a 314 Clipper in use by the US Navy suffered a double engine failure and successfully landed at sea 650 miles east of the Hawaiian island of Oahu. After unsuccessful attempts to repair the engines at sea, the navy attempted to tow the aircraft to land but it was damaged in the process and scuttled. Considering the accumulation of flying hours was probably less than 50,000, this double engine failure rate would be considered totally unacceptable today.

1938: First Pressurised Passenger Aircraft.

It had long been recognized that it was more efficient for aircraft to fly at higher altitudes; aircraft drag reduces in proportion to the associated reduction in air density. Aircraft can also fly in less turbulent air, improving passenger comfort. Unfortunately, there is an associated reduction in air pressure and oxygen levels, which makes flying at higher altitudes uncomfortable for passengers. To overcome these adverse effects it is possible to pressurise the cabin with compressed air from the engines. Military aircraft were able to fly at high altitudes as flight crew were provided with heated flight suits and oxygen masks; however, this was not suitable for fare paying passengers. The selection of a suitable internal cabin pressure is one of compromise between what is comfortable for the passengers, and compatible with the fuselage design.

The Boeing 307 Stratoliner was the first commercial airliner with cabin pressurisation. It first flew in 1938 and entered service in 1940. The cabin was maintained at a pressure equivalent of 8,000 ft altitude, i.e. 11 pounds per square inch (psi). At the cruise altitude of 20,000 ft the pressure outside the aircraft was 6.75 psi, hence there was a differential pressure from

Fig 29: Boeing 307 Stratoliner

inside to outside the cabin of 3.25 psi. This differential pressure imposes a stress cycle on the fuselage every flight, and to accommodate this it was necessary to adopt a cylindrical fuselage design with thicker skins.

The aircraft had accommodation for 33 passengers and five crew, and was designed for use on transcontinental routes with a range of 1,750 miles. Power was provided by four Wright GR1820-G102, 9-cylinder radial engines, each delivering 1,100HP. It reduced the transcontinental flying time by two hours relative to the Douglas DC-3.

Post-WW2 Transatlantic Flights

The intervention of WW2 frustrated the development of intercontinental air travel. In 1942 the British and American governments agreed that American aircraft manufacturers would develop large troop-carrying aircraft, whilst the British concentrated on fighters and bombers. This, together with the need for long range bombers, resulted in the development of larger aircraft with more powerful double-row radial piston engines. In addition, the first turbojet engines were developed for use in fighter aircraft.

Whilst the Stratoliner was undergoing development, Lockheed had been designing a pressurised aircraft called Excalibur. In 1939 Trans World Airlines (TWA), at the instigation of the legendary Howard Hughes, their major stockholder, requested a 40-passenger transcontinental airliner with 3,500 miles range. This was beyond the capabilities of the Excalibur design. TWA's requirements led to the design of L-049 Constellation; however, their initial order was converted to military transport use and designated the C-69.

After WW2 the Constellation came into its own as a fast civil airliner. Aircraft already in production for the USAAF as C-69 transports were finished as civil airliners, with TWA receiving the first on 1 October 1945. TWA's first transatlantic proving flight departed from Washington DC on 3 December 1945, arriving in Paris on 4 December, via Gander and Shannon. TWA's transatlantic service started on 6 February 1946 with a Constellation flying from New York to Paris. The aircraft was later extended in length to increase capacity and range and designated the L-1049 Super Constellation, which first entered service in 1951 with Eastern Airlines.

The aircraft was further modified and designated L-1649 Starliner. Airline service began on 1 June 1957 with a TWA flight from New York to London.

Fig 30: Lockheed L-1649 Constellation

The initial Constellation L-049 was powered by four Wright Cyclone R-3350-745C-18BA, 18-cylinder (twin-row) piston engines, each rated at 2,200 HP. Cruising speed was 275 mph and maximum payload range was 2,290 miles. The Super Constellation L-1049 engines were uprated to Wright Cyclone R3350-972TC-18DA, rated at 3,400 HP each, and had a cruising speed of 305 mph and a max payload range of 4,140 miles. The Starliner L-1649 engines were further modified to R3350-988TC-18EA, which also had a power rating of 3,400HP, a cruising speed of 290 mph and a maximum payload range of 4,940 miles. Seating capacities were 44 for the L-049 and 99 for the Super Constellation and Starliner. The Starliner could reach any European capitol non-stop from any major airport in the US. It was the fastest piston engine powered airliner at ranges over 4,000 miles (6,437 km) ever built.

The Wright R3350 Duplex Cyclone was used to power a number of aircraft, including the Boeing B-29 Superfortress, a high altitude, long range Bomber, of which a total of 3,970 were built; and the Boeing Stratofreighter, the forerunner of the Stratocruiser airliner.

Due to its complexity, the engine took a long time to mature. Early engines suffered from overheating, particularly of the rear row of cylinders; the magnesium crankcase was prone to catch fire, occasionally with catastrophic consequences; carburation problems were also an issue and engine overhaul times were as low as 100 hours. However, its

Fig 31:Boeing B-29 Superfortess

extensive use in military applications and the incorporation of many modifications resulted in an engine that, by the end of WW2, had matured sufficiently to be selected to power the Lockheed Constellation transcontinental airliner. The engine was upgraded to increase horsepower and fuel efficiency, using a process called turbo-compounding, which uses the high velocity exhaust gases to drive a turbine that transfers energy to the crankshaft. Wright was the only aircraft engine manufacturer to put a turbo-compound engine into production.

Turbo-compounding of the Wright 3350 was achieved by placing three power recovery turbines (PRTs) spaced at 120-degree intervals around the rear of the engine. The addition of these between the power and supercharger sections added only 11 extra inches to the overall length, compared to a non-turbo-compounded engine. Each PRT was driven the by the exhaust gases of three front row and three rear row cylinders. The power was transferred to the engine crankshaft through a fluid coupling. Turbo- compounding added about 550 horsepower at take-off power and 240 horsepower at cruise settings over a similar non-turbo-compounded R-3350. These power increases were achieved with a weight penalty of about 500 pounds. Turbo-compounded versions were used to power the Lockheed Constellation Starliner and Douglas DC-7C airliners. Overhaul periods of 3,500 hours were eventually achieved. In Fig 32 one of the three recovery turbines can be recognised by the red blading in the foreground.

Fig 32: Wright R-3350 Turbo-Compound Engine

Boeing introduced the Stratocruiser (Fig 33), an aircraft with a pressurised cabin, into service with Pan Am in 1949, flying between San Francisco, California and Honolulu. The Stratocruiser was a civil airliner version of the military transport Stratofreighter. It was the first double deck passenger aircraft ever to be made and had accommodation for between 55 and 100 passengers, or 28 sleeping berths and five seats. All versions featured a lower-deck lounge and bar, an innovation which, combined with a long range and high speed, set a new standard for luxurious air travel.

The aircraft was larger than the Lockheed Constellation and had a cruising speed of 340 mph at 25,000 ft, powered by four Pratt & Whitney Wasp Major, 28-cylinder engines, delivering 3,500 HP each. The British Overseas Airways Corporation (BOAC) took delivery of its first aircraft in the same year as Pan Am, and began transatlantic services in December of 1949.

In a quest for even more power, Pratt & Whitney introduced the R-4360 Wasp Major, a 28-cylinder, four-row, supercharged radial. The design was ingenious as each row of pistons was slightly offset from the previous row, forming a semi-helical arrangement to facilitate the efficient

Fig 33: Boeing 377 Stratocruiser

airflow cooling of the successive rows of cylinders. The capacity was 4,362 cubic inches (71.5 litres), hence the model designation.

Post WW2, the engine was extensively used by the US military to power a number of heavyweight aircraft, including the B-50 Superfortress bomber, K-97 Stratofreighter, Douglas C-74 Globemaster. Its major civil applications were in the Boeing 377 Stratocruiser and Lockheed Super Constellation transcontinental aircraft. Although fairly reliable, the engines developed an unenviable reputation for in-flight fires, particularly in its Boeing Stratocruiser application, and in addition the Wasp Major was maintenance intensive. Improper starting techniques could foul all 56 spark plugs, requiring hours to clean or replace. The time between overhauls was as low as 600 hours when used in commercial service. One Pan Am aircraft had to make an emergency landing at sea off Hawaii due to a double engine failure in flight.

The Wright Cyclone engines also had poor reliability; at one stage the Lockheed Constellation was jokingly referred to as 'the best three engine aircraft in service'.

The Pratt and Whitney 4360 was the last major development of high powered piston engines, which were eventually superseded by the turbojet and turboprop. Increases in power rating and the associated

Fig 34: Pratt & Whitney R-4360 Double Wasp
28-Cylinder Radial Engine

complexity had an adverse effect on reliability, making them unsuitable for long range twin-engine applications.

The chart below (Fig 35), based on data from an ICAO review in 1953, shows the effect of increases in horsepower on engine failure rates for piston engines. This poor level of reliability was the basis for the introduction of the 60-minute rule for twin-engine aircraft in that year.

The increases in horsepower were broadly in line with the increase engine capacity, which was achieved primarily by increasing the number of cylinders. This inevitably resulted in increased complexity with an associated reduction in reliability. At 3,500 HP the radial piston engine was approaching the limit of its development. An in-flight shutdown rate approaching one/1,000 hours mitigated against the use of these larger engines on twin-engine aircraft. This was all changed by the introduction of the jet engine.

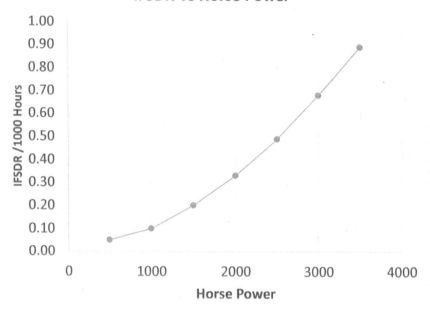

Fig 35: Shut Down Rates/1,000 Hours, Versus Engine Horsepower, for Pre-1953 Piston Engines

1953: Introduction of the 60-Minute Rule

In 1953, in response to the poor piston engine reliability as indicated above, the US Civil Aeronautics Authority, introduced the *60-minute rule* for twin-engine aircraft. This rule stated that, "The flight path of twin-engine aircraft should not be farther than 60 minutes of flying time from an adequate airport". This forced these aircraft, on certain routes, to fly a dog-leg path to stay within regulations, and they were totally excluded from certain routes due to lack of en route airports. The *60-minute rule* was also called the *'60-minute diversion period'*. The totally excluded area was called the *exclusion zone* (see Chapter 4).

As a result of the decision by the British Government in 1942 to exclude itself from the development of large troop carrying aircraft, and its lack of engines of adequate HP at the end of WW2, there were no British aircraft that could compete with the American manufacturers. Consequently, the Lockheed Constellation, Boeing Stratocruiser and Douglas DC-7, another competitor in this market, had a monopoly of intercontinental travel until the emergence of turbojet powered airliners in 1958.

Chapter 3

The Early Military Jet Engine Era

Centrifugal-flow Compressors

Even before the start of WW2, engineers were beginning to realise that engines driving propellers were self-limiting in terms of the maximum performance which could be attained. Propeller efficiency declines as the blade tips approached the speed of sound. If aircraft performance were ever to increase beyond such a barrier, a way would have to be found to use a different propulsion mechanism, and this was the motivation behind the development of the gas turbine (Jet) engine. A young RAF pilot, Frank Whittle, had recognised this as early as 1928, when he wrote a paper on the future of aircraft propulsion. Regrettably, the hierarchy at the British Air Ministry lacked the vision to see the potential of his ideas. They rejected his proposal for a turbojet engine, concluding that it would lack sufficient power. Undeterred, in 1930 Whittle filed patents covering his proposals for jet propulsion and began to design and manufacture a prototype turbojet engine. Whittle set up his own company, Power Jets, but due to lack of funding it was 1937 before his engine became the first jet propulsion engine to be bench tested. It successfully demonstrated the feasibility of the turbojet.

The basic thermodynamic cycle of a turbojet is similar to that of the internal combustion engine: induction, compression, ignition, exhaust. However, in the piston engine combustion occurs at constant volume whilst in the gas turbine engine it occurs at constant pressure. Both the piston engine and the jet engine work on the principle of propelling the aircraft forward by thrusting a mass of air rearwards; the propeller in the form of a large airstream at a comparatively low velocity, and turbojet in the form of a jet of gas at very high velocity.

The propulsive efficiency of this process is the measurement of the ability of the engine to convert the kinetic energy so produced into forward motion of the aircraft. At speeds up to approximately 400 mph the propeller has a higher efficiency than the turbojet, however at speeds above 350 mph the propeller efficiency drops rapidly due to the disturbance of the airflow by the high blade tip speeds. The turbojet is therefore more suited to high speed operation than the piston engine.

Frank Whittle recognised the potential benefits of the turbojet engine in the 1920s. His first engine was the simplest form of turbojet, comprising a single spool with a centrifugal-flow compressor, driven by a single-stage uncooled turbine (Figs 36 and 37). The prototype engine demonstrated 850 lbf of thrust. The Air Ministry was sufficiently impressed to release funding for a flightworthy engine of increased thrust, a model that was designated the Whittle W1.

Maximum thrust was 1,032 lbf at a turbine entry temperature (TET) of 780 degrees centigrade. The air mass flow was 25.4 lb/second and overall pressure ratio (OPR) was 3.8:1. Specific fuel consumption (SFC) was 1.358 lb of fuel per lb of thrust per hour and the engine dry weight was 700 lb.

Funding was also released for the design of a flying test bed, the contract being given to the Gloster Aircraft Company. On 15 May 1941, Gerry Sayer took off in the resulting Gloster E28/39 Pioneer, which had been designed specifically to test fly Whittle's W1 engine. It propelled the aircraft to a maximum speed of 338 mph.

Whittle continued to develop the engine but manufacturing was given to the Rover Company, which resulted in further delays. With an uprated W2 engine, the Gloster E28/29 eventually reached a speed of 488 mph.

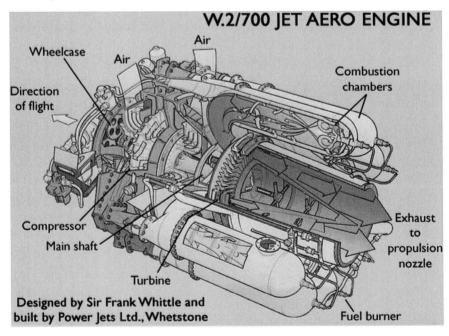

Fig 36: Whittle W1 Turbojet Cross Section (Courtesy of PP Animations)

Fig 37: Whittle W1 Engine

Eventually Rolls-Royce took over further development of the W series engine. It was renamed the Rolls-Royce Welland and used to power the World's first operational jet fighter, the twin-engine Gloster Meteor F1, which first flew on 5 March 1943 and entered operational service in July 1944.

Fig 38: Gloster E28/39

Fig 39: Gloster Meteor

Late in 1945, two F-3 Meteors were modified for an attempt on the world air speed record. This was achieved at Herne Bay in Kent, UK, when Group Captain Hugh Wilson set the first official air speed record by a jet aircraft of 606 mph. The Meteor was initially used to intercept the V-1 Flying Bomb, but despite the best efforts of its pilots it never had the chance to prove itself against the Luftwaffe. Later aircraft were powered by the RR Derwent which had greater thrust, and modifications to improve durability. The Derwent Mk 1 had a thrust of 2,000 lbf, and this was progressively increased to 3,600 lbf in the Derwent Mk 8.

Frank Whittle's struggle to obtain funding inevitably delayed progress of the design and development of his prototype engine. In the meantime, a German physicist and aerodynamic student, Hans von Ohain, was investigating jet propulsion and in 1936 he applied for a patent for a 'Process and Apparatus for Producing Airstreams for Propelling Airplanes'. Whittle's patent expired in 1935 and he was unable to renew it due to lack of funding. It seems highly likely that Von Ohain was aware of Whittle's patent when formulating his ideas. His first design was not a true turbojet but a self-contained engine using the energy from its own power turbine to turn the intake air compressor. It used an electric motor to power the engine compressor. When he attempted to run a prototype he could not control the combustion process, flame shot out in the reverse direction destroying the electric motor.

Despite these early problems, German aircraft designer Ernst Heinkel was sufficiently interested to provide funding to develop the concept. The result was an engine designated HS-3B that went on to power the

Fig 40: Replica of Heinkel HE- 178 (Rostock-Laag Arrival Hall)

Heinkel HE 178, the first jet- propelled aircraft, which made its maiden flight on 27 August 1939 in great secrecy. The engine delivered a maximum thrust of 992 lbf, and propelled the aircraft to a maximum speed of 380 mph, with a range of 125 miles. The historic flight of the HE 178 proved little beyond the feasibility of jet aircraft and luckily for Britain, its engine design was abandoned after Heinkel encountered his own share of official indifference. Luftwaffe officials who witnessed the plane flying told him his turbojet was not needed, believing that they could win the war with piston engine powered aircraft.

Helmut Schelp, the German Ministry of Aviation's director of jet development, gave full details of the research to BMW and Junkers, who began developing their own engines. By now von Ohain was increasingly sidelined, Heinkel's turbojets retained little of his original concept, and none of his designs went into production.

In 1939 the German manufacturer Junkers commenced design of a jet engine which differed from the Whittle engine in that it incorporated an 8-stage axial-flow compressor.

The engine known as the Junkers Jumo 004 became the power plant for the twin-engine Messerschmitt 262, which first flew about 9 months before the Gloster Meteor, making it the World's first operational jet fighter aircraft.

The jet engine had little influence on the outcome of the war; by the time jets were in operational use it was "too little, too late" for both nations. Hans von Ohain later said, "If the British experts had had the vision to

Fig 41: Junkers Jumo 004 Axial Flow Compressor

Fig 42: Messerschmitt ME 262

back Whittle, World War II would probably never have happened. Hitler would have doubted the Luftwaffe's ability to win."

On a mission to Britain in March 1941, General Henry Arnold, US Army Air Corps chief of staff, was amazed to learn that Whittle's engine would soon fly. With America now in the war and providing men and materiel vital to the conflict, Britain was providing scientific know-how in return. All Whittle's research was handed over to the Americans and Britain's early lead in what would become one of the great post-war industries was simply surrendered. Arnold arranged for all the Whittle W1X engineering drawings and some Power Jet's engineers to be flown to the US to help the rapid start of America's jet program. As a result, General Electric produced the J31 turbojet, which was very similar to Whittle W1, producing the same thrust at the same dry weight. The production of this engine kick-started US jet propulsion technology, ultimately to the detriment of its originators.

Fig 43: General Electric J31 Version of the Whittle W1

The J31 powered the Bell P-59A, Airacomet, which was the first American jet fighter aircraft and flew for the first time in October 1942; 18 months after the first flight of the Whittle W1 powered Gloster E28/39.

Further development of the Whittle concept resulted in a change from reverse flow to straight through flow combustion chambers, enabling a simpler arrangement and improved reliability. The straight through arrangement, although increasing length and weight, enabled increases in the diameter of the compressor impellor without excessively increasing the overall diameter of the engine, thereby permitting increases in mass flow and therefore thrust. The first engine to appear with this configuration was the De Havilland Goblin, which first flew in March 1943 in a modified Gloster Meteor, and in September of that year in the De Havilland Venom.

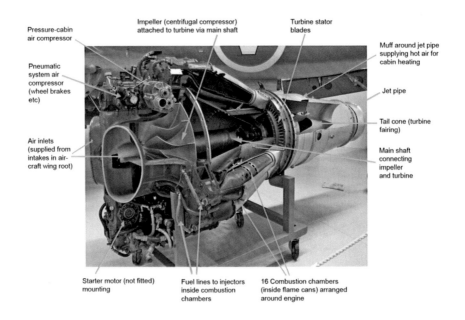

Fig 44: De Havilland Goblin

The Goblin had a thrust of 3,000 lbf at 10,200 rpm. An engine was sent to the US in 1943 where it was used to power the prototype Lockheed P-80 Shooting Star, which had been designed around it, and first flew in January 1944. The P-80 was seen as the answer to the Messerschmitt 242 and the project was treated with great secrecy; so much so that the British engineer who delivered the Goblin engine was detained by the police because Lockheed officials could not vouch for him. After the engine had been mated to the airframe, foreign object damage during the first run-up destroyed the engine. The only replacement engine available was in the prototype De Havilland Vampire, which delayed the first flight until it could be delivered from Britain.

Rolls-Royce had also recognised the advantages of the straight through combustor arrangement and introduced the RR Derwent late in 1943. The Derwent had a double-sided centrifugal compressor and single-stage turbine. The first production version delivered 2,000 lbf thrust at 16,600 rpm and a TET of 850 degrees centigrade. It quickly replaced the Welland in the Gloster Meteor and proved to be a more reliable engine. Further upgrades enabled the engine to deliver 2,400 lbf thrust. General Electric also developed a similar engine, designated the I-40, delivering 4,000 lbf thrust, but they were forced to relinquish production of the I-40 to Allison who renamed it the Allison J33. The engine had better performance than the De Havilland Goblin and was selected as the power plant for production versions of the Lockheed P-80.

Initially the P-80 Shooting Star was notoriously difficult to fly and resulted in a number of accidents. Delays in production prevented active service during WW2, although 1,714 were produced before production ceased in 1950. AP-80 became the first jet aircraft to complete a transcontinental flight, when on 27 January 1946 Colonel William H. Councill flew non-stop across the US between Long Beach and New York, a distance of 2,457 miles, in 4 hrs 13 min at an average speed of 584 mph.

Fig 45: Lockheed P-80 Shooting Star

Towards the end of the war, a series of increasingly more powerful engines emerged from Rolls-Royce and the company went on to become arguably the pre-eminent manufacturer of turbojets.

In 1944 Rolls-Royce produced its third jet engine, the Nene, which was basically a scaled-up version of the Derwent. The engine was designed and built in a short five-month period and had its first run in October of that year. The engine had a thrust of 5,000 lbf at 12,300 rpm and weighed 1,600lb. Its two main applications were in single seat fighters, the Hawker Sea Hawk and Supermarine Attacker, of which 542 and 182 were built respectively.

The Nene is probably better known for its part in the success of the Russian MIG-15 fighter aircraft that played a prominent part in the Korean

Fig 46: Rolls-Royce Nene Turbojet

War, 1950-54. The Soviet aviation minister Mikhail Krunichev, and aircraft designer A S Yakovlev, suggested to Premier Joseph Stalin that the USSR buy the fully developed Nene engines from Rolls-Royce for the clandestine purpose of copying them. Stalin is said to have asked, "What fool will sell us his secrets?" However, in 1946, before the Cold War had really begun, the British Labour Government, under the Prime Minister Clement Attlee, was keen to improve diplomatic relations with the Soviet Union and was receptive to the idea. Accordingly, as a gesture of good will, Attlee authorised Rolls-Royce to export 25 Nene engines to the Soviet Union, with the reservation that they were not to be used for military purposes.

The Russians immediately reneged on the agreement and back-engineered the engine to produce the Klimov RD-45, which was subsequently modified to produce the VK-1. It is said that when Russian diplomats visited the British engine manufacturing facilities they wore soft soled shoes so that swarf from the manufacturing processes would be embedded therein and later analysed to identify the material composition.

The VK-1 engine was used to power the successful MIG-15 Jet fighter, of which about 12,000 were made in Russia and approximately a further 6,000 made under licence in Poland and Czechoslovakia.

The MIG-15 was produced in time to be deployed against United Nations (UN) forces in North Korea in 1950, causing the loss of several

Fig 47: Klimov VK-1 Engine

B-29 bombers. The UN cancelled their daylight bombing missions, much to the consternation of the Americans, who later referred to it as the 'Nene Blunder'. On discovering that the Nene had been copied without any licence agreement, Rolls-Royce attempted to claim £207m in licence fees, but not surprisingly were unsuccessful.

Pratt & Whitney also built the engine under licence and designated it the Pratt & Whitney J42. It was used to power the Grumman F9F-2 fighter

Fig 48: USSR MIG-15 Fighter

which also fought in the Korean War. Pratt & Whitney went on to produce the J48, an uprated version of 7,250 lbf thrust, based upon the RR Tay 44 engine, which it also built under licence. The J48 was fitted to later versions of the Grumman F9F-2. The J42 and 48 were the first Pratt & Whitney engines to enter production and formed the bases for their further engine developments.

Hispano Suiza also built the Nene under licence and later developed its own version, the Verdon, which delivered 7,700 lbf thrust, making it the highest rated centrifugal-flow compressor turbojet ever produced. The Verdon powered the French Dassault Mystere 1V Fighter Bomber, which first flew in 1952.

In a similar timescale De Havilland produced the DH Ghost, a scaled-up version of their Goblin engine with similar performance to the RR Nene; this also ran for the first time in 1944. The Ghost Mk1 was slightly larger and heavier than the Nene, but produced a similar thrust of 5,000 lbf at 10,250 rpm. The engine was famously used to power the ill-fated DH Comet 1 Jetliner (See Chapter 4).

Table 1 (Appendix 2) summarises the approximate performance details of the several engines which had similar geometric configurations, being single spool, centrifugal-flow compressors with straight through combustors and single stage turbines, all of which were derived from the original Whittle engine.

By 1950, jet aircraft of all the combatant nations were powered by developments of British designs based on Whittle's original concept. At 7,000 lbf thrust the centrifugal-flow compressor arrangement was approaching the limit of its growth capability and further increases in diameter and mass flow became prohibitive, due to the adverse effect of increased drag. By the time the higher-rated versions of the centrifugal-flow configurations were becoming operational, designers were already pursuing the smaller diameter alternative of the axial-flow compressor arrangement, which offered greater potential for thrust growth and increased thermal efficiency.

The Single-Spool Axial-Flow Compressor

As described above the Junkers Jumo 004 was the first engine to use an axial-flow compressor. The Jumo 004 had a thrust rating of 1984 lbf, at 8,700 rpm. The air mass flow was 55 lb/second, OPR was a modest 3.14:1, TET was 937deg centigrade and SFC 1.39. The engine was 32 inch in diameter, 152 inch in length and weighed 1,640 lb, giving a thrust-to-weight ratio of

1.2:1. Overhaul life is believed to have been less than 50 hours.

Frank Whittle had considered the relative merits of axial versus centrifugal compressors before embarking his jet engine design. It was his opinion that, at the time, the axial compressor was too fragile, with too many potential problems, so he opted for the more robust and less risky design of the centrifugal compressor. The successful experience with the derivatives of his engine vindicates his decision. It was the right one for that very early stage of jet engine development. By the time it became inevitable that a change to axial-flow compression was required to facilitate further increase in thrust growth and engine efficiency, many of the problems with the rest of the turbo-machinery were understood.

It is interesting to compare the specification of the RR Derwent 1 centrifugal-flow, 2,000 lbf thrust engine, with that of the Jumo 004. The Derwent had better SFC, significantly better thrust-to-weight ratio and lower TET. The only disadvantage of the Derwent was the higher drag due to its larger diameter.

In addition to the Whittle engine data provided by the British government, US company General Electric also had available to them German technology obtained post WW2 in 1945 as part of Operation Lusty (**Lu**ftwaffe **S**ecret Technolog**Y**), no doubt this would have included details of the Jumo 004. In parallel with designing the Allison/GE J33, GE also embarked on the design of the J35, an axial-flow design using an 11-stage axial compressor and single-stage turbine. The engine delivered 5,600 lbf thrust at 8,000 rpm and was America's first axial-flow compressor engine. The engine powered the Republic F-84 Thunderjet on its first flight in 1946, and late in 1947, like the J33, complete responsibility for the development and production of the engine was transferred to Allison. The engine was

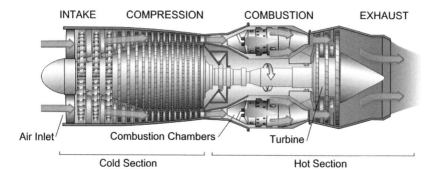

Fig 49: Layout of a Single-Spool Axial-Flow Turbojet

used in other applications, the most notable of which was the Northrop F-89 Scorpion. A total of 7,524 Thunderjets and 1,050 Scorpions were built.

GE quickly followed the J35 with the J47, an uprated version of 5,970 lbf thrust; this had a 12-stage HP Compressor which enabled a 5 percent increase in pressure ratio and an 8 percent reduction in SFC. The engine first ran in 1947 and entered service in 1949 powering the North American F-86 Sabre. It was successfully used in a number of military aircraft and a total of 36,500 were built. Initial overhaul life was very low at 15 hours, but this was increased to approximately 625 hours in 1956. Installed on the F-86F Sabre it reputedly achieved an IFSDR of one/33,000 hours during 1955 and 1956. Despite this excellent reliability the engine was not used in any civil applications. The engine was further uprated to 9,500 lbf thrust for use in the F 86-H Sabre and YF 84J Thunderstreak.

In 1944 Rolls-Royce had started designing the Avon, an axial-flow gas turbine, intended as a future replacement for the Nene. The Avon design team was headed by Cyril Lovesey, who had previously been in charge of the RR Merlin development. The first engines were designated AJ 65, an abbreviation for *Axial Jet 6,500 lbf,* and were later renamed the Avon RA series. Development started in 1945 and the first prototypes were built in 1947. Introduction was somewhat slowed by a number of minor problems, however the engine eventually entered production in 1950, the original

Fig 50 Martin B-57 Canberra Bomber

Fig 51: Rolls-Royce Avon

RA.3/Mk.101 version providing 6,500 lbf (29 kN) thrust in the English Electric twin-engine Canberra B2 Bomber, which entered service in 1951.

The Canberra could fly at a higher altitude than any other bomber through the 1950s, setting a world altitude record of 70,310 ft (21,430 m) in 1957. On 21 February 1951, an RAF Canberra B Mk2 became the first jet powered aircraft to complete a transatlantic crossing without refuelling, flying from Aldergrove, Northern Ireland, to Gander, Newfoundland. The flight covered almost 1,800 nautical miles (3,300 km) in 4hr 37min. The aircraft was being flown to the USA to act as a pattern aircraft for the Martin B-57 Canberra, which was to be built under licence by the Glenn L Martin Company and used by the US Air Force in the Korean War.

To expedite production, the first B-57As were largely identical to the Canberra B2, with the exception that the more powerful Armstrong Siddley Sapphire engines (Fig 52) of 7,200 lbf (32 kN) thrust were used instead of Rolls-Royce Avon. The engines were also built under licence in the US by Curtiss Wright and known as the Wright J65.

The Canberra was the victorious aircraft flown in 'The Last Great Air Race' in 1953, which took place from London Heathrow to Christchurch International Airport New Zealand, a distance of 12,300 miles (19,800 km). The winner arrived 41 minutes ahead of its closest rival after a flight of 23hr 51min; to this day the record has never been broken.

There were many variants of the Avon, the most notable of which were the 7,500 lbf thrust RA7 that powered the French Mystere fighter and later versions of the Canberra; the 11,000 lbf thrust RA 29 that powered the De Havilland Comet 4, and which became the first jet-propelled passenger aircraft to complete a scheduled transatlantic crossing; and the French Sud Aviation Caravelle, the first twin-engine jet airliner. Early Avons had a 12-stage compressor and a 2-stage turbine, the RA 29 had a 16-stage compressor and a 3-stage turbine.

Another single-spool, axial-flow compressor turbojet of note was the Armstrong Siddeley Sapphire which first flew in 1951 in a Hawker Hunter, being later replaced by the Rolls-Royce Avon. Other military applications included the Gloster Javelin, and Handley Page Victor; there were no civil applications.

At the end of WW2, the French company SNECMA obtained details of the German BMW 018 axial-flow design, and from it produced their ATAR 101 engine, which had its first flight in December 1951 in the Dassault Ouregan M.D. 450. This was the first French fighter-bomber and it played a key role in the resurgence of the French aviation industry, which had been decimated by the German occupation of WW2. All its other applications were in French military aircraft, which included the

Fig 52: Wright J65, Licensed version of Armstrong Siddeley Sapphire

Dassault Mirage, Super Mystère and Étendard II.

General Electric went on to produce the J79 which was based on the J73 but with a 17-stage compressor and 3-stage turbine, which enabled it to operate at a pressure ratio of 13.5:1 and deliver 11,900 lbf thrust.

Applications were primarily military, however it was designated the CJ805-3 for civilian applications and used to power the Convair 880 four-engine narrow-body jetliner.

Fig 53: General Electric J79 Axial-Flow 17-stage compressor

The data in Table 2 (Appendix 2) illustrates the beneficial effects of increased mass flow on thrust-to-weight ratio and increased pressure ratio on SFC.

Adding more compressor stages to the single-spool rotor was proving increasingly more difficult; in addition to the mechanical complexity of adding more stages, to optimize aerodynamics the front end of the compressor ideally wants to rotate slower than the rear end. The next logical step in the development was therefore the twin-spool turbojet, with two axial-flow compressors each operating at their optimum speed.

The Twin-Spool Axial-flow Compressor Turbojet

The Bristol Aero Engines, BE.10 Olympus, was the world's first twin-spool axial-flow turbojet aircraft engine. It was developed to deliver 11,000 lbf thrust and had a 6-stage low pressure compressor (LPC) driven by a single-stage turbine, and an 8-stage high pressure compressor (HPC) also driven by a single-stage turbine. First running in 1950, its initial use was as the power plant of the Avro Vulcan Delta Wing Bomber which entered service with the RAF in 1956. The design was further developed to the MK 320, delivering a thrust of 19,000 lbf for supersonic performance in the British Aircraft Corporation TSR-2. However, the aircraft project was cancelled before it entered production.

The MK 320 was further developed to become the Rolls Royce/SNECMA Olympus 593, and was used as the power plant for the Concorde Supersonic Transport. Versions of the Olympus 101 engine were licensed to Curtiss Wright in the USA as the TJ-32 or J67 (military designation) and the TJ-38 'Zephyr'.

Unlike General Electric and Rolls-Royce, Pratt & Whitney did not pursue a single-spool axial compressor design, preferring to concentrate their efforts on a 2-spool configuration.

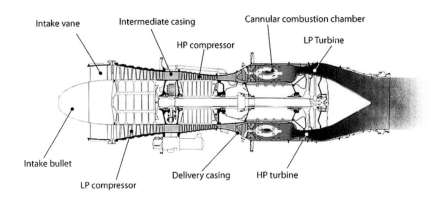

Fig 54: Bristol Olympus Mk 101

The J57 was developed as a military engine and was used to power the Convair F-106 Delta Dart (Fig 56), which first flew in December 1956, and the North American F-105 Thunderchief (Fig 57), which flew in May 1956.

The layout of the J57 was similar to that of the Bristol Olympus 101, with the exception that the LP Compressor had nine stages driven by a 2-stage LP Turbine, and the HP Compressor had seven stages driven by a single-stage H P Turbine. This engine was designated the Pratt & Whiney JT3 in its civil application and became a very successful engine.

The military requirements for successive conflicts had a profound effect on the rate of development of gas turbine engines, to the ultimate benefit of the civil aircraft industry. Many of the post-war passenger aircraft were powered by engines that had been proven in military applications.

Fig 55: Pratt & Whitney J57 (JT 3) Twin-Spool Turbojet

Fig 56: Republic F-105 Thunderchief

Fig 57: Convair F-106 Delta Dart

Chapter 4

The Civil Jet Engine Powered Aircraft Pre-ETOPS Era

Four-Engine Airliners

In 1942 during WW2, the USA and the United Kingdom agreed to split responsibility for building multi-engine aircraft types for British use: the US would concentrate on transport aircraft while the UK would concentrate on their heavy bombers. It was soon recognized that as a result of that decision the UK would be left at the close of the war with little experience in the design, manufacture and final assembly of large transport aircraft, nor would there be the infrastructure and trained personnel to undertake the task should the UK wish to do so. On the US side, the massive infrastructure created during the war would allow them to produce civilian aircraft based upon military transport designs; and crucially these would have to be purchased by the UK, Empire and Commonwealth to meet their post-war civilian transport aviation needs. This, of course, is what happened with the Lockheed Constellation and Boeing Stratocruiser as mentioned in Chapter 2.

A committee was set up in February 1943, under the leadership of Lord Brabazon, to investigate the future needs of the British civilian airliner market. Committee members included Geoffrey De Havilland of the De Havilland Aircraft Company Ltd; Alan C Campbell-Orde, assistant to the Chairman of the British Overseas Airways Corporation (BOAC); Sir William P Hildred, Director General of Civil Aviation, and representatives from the Air Ministry, Ministry of Aviation, Ministry of Aircraft Production and the Air Registration Board. They studied a number of designs and technical considerations, meeting several times over the next two years to clarify the needs of different market segments. The final report initially called for the construction of four general designs that had evolved from the committee's deliberations, a fifth design was added later.

As previously mentioned (Chapter 3), De Havilland was producing its own jet engines, the DH Goblin and its derivative the DH Ghost. De

Havilland also had experience of designing civil aircraft of moderate size and could see the potential offered by a jet-propelled airliner.

The formation of the Brabazon Committee represented unusually forward thinking, considering the wartime situation at the time. One of its recommendations was for a pressurised, transatlantic mail plane, that could carry one ton of payload at a cruising speed of 400 mph. Committee member Sir Geoffrey De Havilland, head of the De Havilland Aircraft Company, used his personal influence and his company's expertise with military jet-propelled aircraft to specify a turbojet-powered design. The committee accepted the proposal, calling it the 'Type IV' (of five designs), and in February 1945 it awarded a production contract to De Havilland under the designation Type 106. Initial design of the DH 106 focused on short and intermediate range mail planes with a small passenger compartment with as few as six seats; this was later redefined as a medium-range airliner with a capacity of 44 seats. Out of all the Brabazon Committee design proposals, the DH 106 was seen as the riskiest, both in terms of introducing untried design elements, and for the financial commitment involved. However, BOAC found the DH 106 specifications attractive, and initially proposed a purchase of 25 aircraft; this number was later reduced to ten in December 1945.

The DH 106, later named the Comet, was a major step forward relative to De Havilland's previous civil aircraft experience. Most of their previous civil aircraft had been small biplanes, with the exception of the DH 95 Flamingo which was a 17-seat high wing monoplane of metal construction that had operated throughout the war. Their only other all-metal aircraft was the DH Dove, an 8-seat twin engine propeller driven short-range aircraft, which entered service in 1946. The Comet was to be the first aircraft of any type to be powered by four jet engines. To take advantage of jet propulsion it would fly higher and faster than any other civil airliner then in service. At the higher altitude, the differential pressure from inside the fuselage to the outside atmosphere would be twice that of current pressurized cabins. To reduce drag at the higher speeds a degree of sweep was included in the wing design, and the engines were positioned in the wing roots close to the fuselage to minimize drag. On 2 May 1952, as part of BOAC's route proving trials, G-ALYP took off on the world's first jet powered flight with fare-paying passengers, inaugurating a scheduled service from London to Johannesburg. Each De Havilland Ghost engine delivered 5,000 lbf thrust enabling the aircraft to cruise at 460 mph at an altitude of 42,000ft.

Fig 58: De Havilland Comet 2

In their first year of operation the Comet 1's carried 30,000 passengers. The Ghost engines allowed the Comet to fly above turbulent weather conditions that competing propeller aircraft had to fly through, they ran smoothly and were less noisy than piston-engine aircraft, had lower maintenance costs, and were fuel-efficient at the higher altitudes. In summer 1953 eight BOAC Comets left London each week on routes to Johannesburg, Tokyo, Singapore and Colombo. The Comet 1 range of 1,500 miles prevented it from operating on transatlantic routes; however, in 1953 plans were already in place to introduce a Comet 2 with an increase in range to 2,600 miles, made possible by the replacement of the DH Ghost engines with Rolls-Royce Avon engines, each delivering 7,000 lbf thrust (see Fig 50). The increase in range was also accompanied by an increase in cruise speed to 490 mph.

Early in its career the Comet 1 experienced two crashes during takeoff due to loss of lift at high angles of attack. This was corrected by a modification to the leading edge profile of the wing. A more serious problem developed when two aircraft mysteriously crashed in 1954, with the result that all Comets were grounded whilst investigations took place. It was eventually established that the pressurisation of the fuselage during each flight cycle had resulted in high peak tensile stresses occurring in rivet holes adjacent to the windows and other apertures in the fuselage. The peak stresses in these rivet holes, superimposed upon the already elevated stress field around the rectangular apertures, were as high as 80 percent of the minimum ultimate tensile strength (UTS) of the aluminium alloy material (DTD 564B) used for making the fuselage skins.

The effects of Low Cycle Fatigue (LCF) were clearly underestimated by the De Havilland design team. The fuselage skins were no thicker than those of the other pressurized aircraft of the day, even though they were exposed to almost twice the pressure loading. De Havilland had recognised the need to test the structural integrity of the fuselage prior to entry in to service, and did this by exposing the fuselage to a one-off proof test of twice the normal operating differential pressure, followed by 18,000 cycles at the normal differential operating pressure of 8.25psi. The test failed to expose the fatigue problem; as a result, De Havilland was confident that the design had the desired level of integrity. As part of the post-failure investigation these tests were repeated but without the prior proof test, producing cracks similar to those that resulted in the failed aircraft.

It is possible that the original over-pressurisation test caused yielding in the rivet holes, which resulted in a residual compressive stress when the pressure was removed, if so this would have given an optimistic result when the subsequent pressure cycles were applied. Additionally, there is a considerable variability in material fatigue strength properties which was probably not fully understood at the time. The assumption that a single test demonstrating 18,000 cycles fatigue life would be representative of a typical service life was therefore over-optimistic.

As a result of these metal fatigue failures, De Havilland redesigned the aircraft as a larger version, designated the Comet 4. It had a strengthened

Fig 59: De Havilland Comet 4

fuselage, increased capacity to 81 seats, an operating range of 3,200 miles and a cruising speed to 520 mph. This was enabled by the availability of uprated Rolls-Royce Avon engines each delivering 10,500 lbf thrust. The Comet 4 entered service on transatlantic routes with BOAC on 30 September 1958, four years after the grounding of the Comet 1. A total of 76 Comet 4s were delivered between 1958 and 1964 when production ceased.

The last two Comet 4C aircraft produced were modified as prototypes to meet a British requirement for a maritime patrol aircraft for the Royal Air Force, initially named Maritime Comet, and later the Hawker Siddeley Nimrod. The Nimrod entered service with the RAF in 1969; five Nimrod variants were produced and a total of 49 production aircraft were built, the final Nimrod aircraft were retired in June 2011.

During the time that De Havilland had been researching the metal fatigue problem, Boeing and McDonald Douglas had been designing their own jet-propelled airliners, the Boeing 707 and Douglas DC-8. Both benefitted greatly from the better understanding of metal fatigue generated by the De Havilland investigation, and were able to incorporate the lessons into their designs. This, together with the data provided on Whittle's gas turbine, and gleaned from the German developments at the end of the war, enabled them to overtake De Havilland in producing

Fig 60: Boeing 707

aircraft of higher capacity, speed and range, ready to enter service at the same time as the Comet 4.

The Boeing 707 and Douglas DC-8 were similar in concept, in that they were each initially powered by four Pratt & Whitney JT3 turbojets (see Fig 54), two mounted under each wing. Like the RR Avon, the Pratt & Whitney JT-3 had an axial-flow compressor, which enabled lower diameter engine cowlings than would have been achievable with a centrifugal-flow compressor. As drag is proportional to velocity squared, it was important to keep this to a minimum at the higher cruising speeds made available by jet propulsion. An additional advantage relative to the Comet 4 was that it was easier to incorporate the increases in engine diameter that emerged with further developments in engine design.

The next major advancement in engine design was the introduction of the low bypass ratio turbofan. Rolls-Royce had been developing their Conway 2-spool, low bypass ratio engine for potential military use. Bypass ratio is the ratio of the air bypassing the engine core versus that flowing through the engine core. Boeing had declared an interest in the engine, as even though it had limited bypass ratio (0.25:1) in keeping with its original in-wing mounting, it would increase the range of the 707-420 by 8 percent compared to the otherwise identical 707-320 powered by the non-bypass Pratt & Whitney JT4A (J75). McDonald Douglas also showed an interest and in May 1956 Trans Continental Airways ordered Conway powered DC-8s, orders for the Conway powered Boeing707 followed from Alitalia and Canadian Pacific Air Lines; BOAC, Air India, Varig and

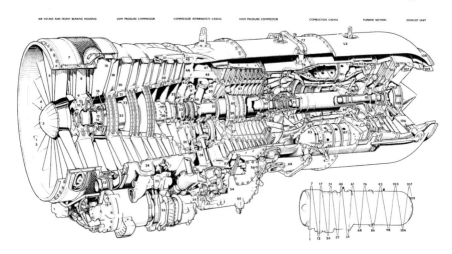

Fig 61: Rolls-Royce Conway RCo-12

Lufthansa. The initial engine was the Conway RCo-10 delivering 16,500 lbf thrust; however, the RCo-10's development was so smooth that after delivering a small number for testing purposes, further deliveries switched to the 17,150 lbf RCo-12, which was designed, built and tested before the airframe testing was complete. These models also featured a distinctive scalloped silencer, and a thrust reverser that could provide up to 50 percent reverse thrust.

Although successful in this role, only 69 Boeing 707's and Douglas DC-8s were built with the Conway, due largely to the delivery of the first US built bypass engines. The RR Conway was the first commercial aero-engine to be awarded clearance to operate for periods up to 10,000 hours between major overhaul, an indication of the improvement in reliability of turbofans relative piston engines.

Pratt & Whitney, aware of the threat posed by the Rolls-Royce Conway, ingeniously developed its JT3C-7 engine into a bypass turbofan by a fairly straightforward modification which could be accomplished in a suitably equipped overhaul shop.

The engine was designated the JT3D-1 and was claimed to offer 13 percent reduced cruise SFC, 33 percent increased takeoff thrust, 15 percent less specific weight (lb/lb of thrust) and a 10 Db lower noise level than the JT3C-7. This was achieved by replacing the first three stages of the LPC with two fan stages to provide a bypass ratio of 1.42:1; an additional LPT stage was added to provide the extra power required to drive the new fan and compressor assembly.

The JT3D-1 was rated at 17,000 lbf thrust and entered service on a Boeing 707-120B with American Airlines in March 1961. The JT3D-1 had

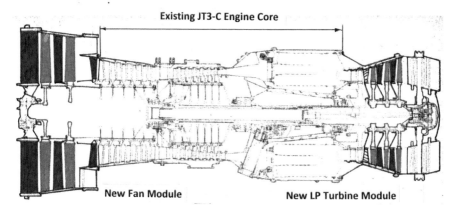

Fig 62: Pratt & Whitney JT3D-1 Modification Package

a lower SFC and specific thrust than the Conway RCo-12 and the configuration had greater growth potential than the Conway as the latter had less scope for further increases in bypass ratio. The engine was later uprated to 19,000 lbf thrust to power the Boeing 707-320B and DC8-63 long range aircraft. The Maximum takeoff weight of these aircraft was twice that of the piston engine powered Lockheed Constellation and Boeing Stratocruiser.

The next major step in civil aircraft and engine design came in 1963 from a requirement from the US Air Force for a large transport aircraft with a load capacity of 180,000 lb, a speed of 500 mph and an unrefuelled range of 5,750 miles. It was required that the aircraft should be powered by four engines each delivering 41,000 lbf thrust, a considerable increase on anything available at the time. In 1965 Lockheed's aircraft design was selected for the new C5 Galaxy transport, the largest military aircraft in the world at the time.

The aircraft was powered by four General Electric TF 39 high bypass ratio turbofans, each having a BPR of 8:1 and delivering 43,000 lbf thrust.

The GE engine concept was revolutionary, as all engines beforehand had a BPR less than 2:1. The high bypass ratio turbofan offered a step change in engine performance, delivering a thrust of 43,000 pounds and a reduction in SFC of about 25 percent relative to low bypass ratio engines. The overall compressor pressure ratio was also increased to 25:1. The first engine went for testing in 1965; between 1968 and 1971 a total of 463 TF39-1 and -1A

Fig 63: Lockheed Galaxy C-5

Fig 64: General Electric TF39 on Lockheed C-5 Galaxy

engines were produced and delivered to power the C-5A fleet.

By the mid-1960s, due to the success of the Boeing 707 and Douglas DC-8, airport congestion was becoming a problem, caused by increasing numbers of passengers carried on relatively small aircraft. To address this, Juan Trippe, president of Pan Am, asked Boeing to build a passenger aircraft more than twice the size of the 707. The availability of the higher thrust, high bypass ratio engine, developed for the Galaxy C5, made it possible to design an all new aircraft to satisfy Pan Am's requirement, the iconic Boeing 747-100 Jumbo Jet which was powered initially by four Pratt & Whitney JT-9D-3A, 5:1 high bypass ratio jet engines, each delivering 44,250 lbf thrust.

The JT9-D3A was a 2-shaft engine consisting of an LP spool with a single-stage 95.6 inch diameter fan and 3-stage LPC (Booster) driven by a 4-stage LPT; and an 11-stage HPC driven by a 2-stage HPT. The BPR was significantly less than the TF39 at 5.2:1. The engine entered service on a Pan Am 747-100 in 1970.

General electric developed the TF 39 into the CF6 civil variant but with a reduced BPR of 5.76:1; the architecture was similar to that of the JT9-3A with the exception that there was only 1 LPC stage but 16 HPC stages producing an OPR of 25:1; it had 5 LPT stages.

Fig 65: Pratt & Whitney JT9D-3

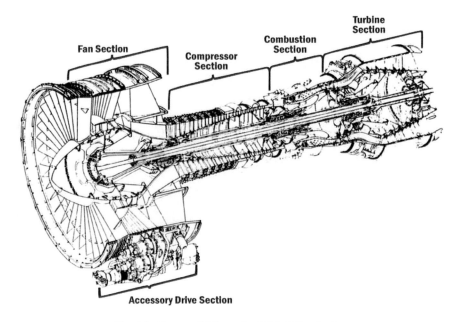

Fig 66: General Electric CF6-6 Turbofan

The initial 747-100 aircraft had a cruising speed of 550 mph at 35,000 ft altitude, and a maximum range of 6,100 miles, with seating for up to 550 single class passengers.

The Boeing 747 entered service on 22 January 1970 with Pan Am flying from New York to London. It became the aircraft of choice for long range travel and by the end of 2014 almost 1,500 aircraft had been delivered to customers. The aircraft was a huge success and rendered the 707 and DC-8 obsolete. It was this aircraft that generated the requirement for engine thrusts exceeding 45,000 lbf, resulting in the high bypass ratio engine configurations that dominate civil aircraft propulsion to day.

Rolls-Royce and General Electric also developed engines for the aircraft. Later versions of the aircraft, such as the 747-400, increased the range to 8,830 miles with seating for up to 660 single class passengers at a slightly higher cruise speed of 570 mph. To achieve this, the engines on offer were gradually uprated to deliver up to 63,000 lbf thrust. The latest version, the 747-8, has a range of 9,200 miles and is powered exclusively by the GEnx-2B67 rated at 66,500 lbf thrust.

Fig 67: Boeing 747-400

Three-Engine Aircraft, Tri-jets

Medium Range

The success of the Boeing 707 and Douglas DC-8 on intercontinental routes resulted in the availability of more efficient low bypass ratio turbofans, opening the door to the design of 3-engine aircraft (Tri-jets), of short to medium range.

For the regional jet market Rolls-Royce embarked on the design the RB-163 Spey, a low bypass ratio turbofan, aimed at the BAC 111 rear mounted twin-jet, and De Havilland (later Hawker Siddeley) Trident tri-jet.

Fig 68: Hawker Siddeley Trident 1C

At the same time Boeing were designing the 727 tri-jet, which was very similar to the Trident, and was considering the Spey engine as the possible powerplant. To improve their chances of securing the order, Rolls-Royce entered into an agreement with the American company Allison to manufacture the engine under licence and designated the Allison AR 963. Pratt & Whitney were at the time conducting a design study for a low bypass ratio turbofan based on its 9,000 lbf J52 (JT8A) 2-spool turbojet that entered service on a US Navy A-6 Intruder in 1963. In spite of the JT8D being approximately 1,000 lb heavier than the Spey, which was further along in its development cycle, both Eastern and United Airlines declared their preference for the Pratt & Whitney engine. Boeing acquiesced to customer preference and the JT8 became the sole power plant on the 727, the first variant being the 727-100 powered by the 14,000 lbf thrust JT8D-1.

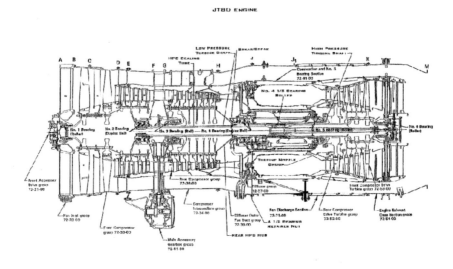

Fig 69: Pratt & Whitney JT8-D

Both the Trident and Boeing 727 went into service in 1964, and coincidently the Federal Aviation Authority (FAA) revised its 60-minute rule to exclude tri-jets; this considerably improved the attractiveness of the Boeing 727 to US Airlines for transcontinental flights. In addition, it later opened the door to the design of larger tri-jets capable of intercontinental travel.

The Boeing 727-100 was powered by three Pratt & Whitney JT8D-1 low BPR engines (1.06:1) each delivering 14,000 lbf thrust; this was later

Fig 70: Boeing 727-200 Tri-jet

uprated to 17,000 lbf to power the 727 Advanced, which had increased range and capacity making it attractive for transatlantic travel. In all 1,832 Boeing 727's were built over a period of 24 years, making it the most successful aircraft of its time.

The Trident failed to penetrate the US market, being eclipsed by the Boeing 727 which was larger and better able to take off from short runways. Its customers were predominately European, Middle East and Asian airlines; a total of only 117 were sold.

Long Range Tri-jets

In 1966, American Airlines specified a requirement for a wide-body aircraft, smaller than the Boeing 747, but capable of flying similar long-range routes from airports with shorter runways. The high bypass ratio engines available at the time were capable of delivering sufficient thrust to make a Tri-jet a suitable proposition. McDonnell Douglas offered the DC10, its first commercial airliner following the merger between McDonnell Aircraft Corporation and Douglas Aircraft Company in 1967. The original aircraft had a maximum range of 3,800 miles with seating for 380 single-class passengers. The engines chosen were General Electric (GE) CF6-6D high BPR turbofans, each delivering 40,000 lbf thrust (see Fig 65). Range was eventually increased to 6,600 miles in the DC10-30 powered by the 51,000 thrust CF6-50C. The first DC10-10 entered commercial service with American Airlines on 5 August 1971, on a round trip flight between Los Angeles and Chicago.

The DC10-30 was the most common model produced; in addition to the General Electric CF6-50 turbofan engines which improved fuel

Fig 71: McDonnell Douglas DC10-30

efficiency, it had larger fuel tanks to increase range. It commenced service on 30 November 1972, with Swissair and KLM as its first customers. A larger version known as the MD 11 entered service in December 1990 with Finnair. The aircraft had seating for 410 single-class passengers, a maximum range of 7,800 miles and was powered by three Pratt & Whitney 4462 series engines. However, the initial aircraft and engine combination suffered from a shortfall in performance, to an extent that caused both American Airlines and Singapore Airlines to cancel their orders. Modifications were incorporated to improve the aircraft performance but this took until 1995, by which time sales had been adversely affected. A longer range version, the MD11-ER, which included these performance improvements, was launched in 1995, but only five passenger versions were sold. Boeing and McDonnell Douglas merged in 1998 and shortly afterwards the production of passenger versions of the aircraft was ceased.

Lockheed also introduced a tri-jet, the Lockheed Tristar (L1011), to compete with the DC-10. The Rolls-Royce RB 211-22B 3-shaft engine was exclusively chosen to power the aircraft.

The RB 211 differed from the GE and Pratt & Whitney high bypass ratio engines in that it utilised three rotating shaft systems as opposed to the 2-shaft configuration of the competing Pratt & Whitney 4000 and GE CF6 series engines. Unlike the American engines Rolls-Royce had no prior military experience with a high bypass ratio turbofan, and as a result of significant development problems with the engine, Rolls-Royce went into liquidation in 1971. The Company was taken over by the UK Government

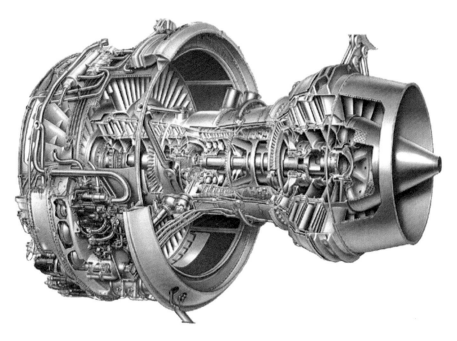

Fig 72: RB211-22B

and was eventually able to meet its commitments to Lockheed, albeit with an initial shortfall in performance. As a result of the engine problems the Tristar entered service later than originally planned.

The first variant to be produced was the L1011-1 which was powered by three RB 211-22B engines each delivering 42,000 lbf thrust. Seating was for a maximum of 400 single-class passengers (20 more than the DC10-10) and maximum payload range was 3,084 miles. The first aircraft was delivered to Eastern Airlines in April 1972, six months later than the DC 10-10. Later variants included the L1011-200 and L1011-500, both of which were powered by RB 211-524 B engines, each delivering 50,000 lbf thrust. The L1011-200 had a range of 4,150 miles with 400 passengers and the L1011-500 had a range of 6,150 miles with 330 passengers. Both of these variants were available one year and three years respectively later than the DC10-30, which adversely affected sales. Lockheed needed to sell 500 airliners to break even, but in 1981 the company announced production would end with delivery of the last L-1011 on order in 1984. A total of 250 Tristars were sold, compared with 446 DC-10s, partly because of the Tristar's delayed introduction but particularly because a larger version with a longer range was not initially offered. Under state ownership costs

Fig 73: Lockheed L1011 Tristar

at Rolls-Royce were tightly controlled, the company's efforts largely went into the original Tristar engines, which needed considerable modifications between the L-1011's first flight and service entry.

The competition, notably General Electric, was very quick to develop their CF6 engine with more thrust, which meant that a heavier intercontinental DC-10-30 could be more quickly brought to market. The flexibility afforded to potential customers by a long-range DC-10 put the L-1011 at a serious disadvantage. In an effort to boost sales and secure the Japanese market, Lockheed had secretly bribed several members of the Japanese government to subsidize a purchase by All Nippon Airways (ANA) of L-1011s; however, this caused an international scandal when the bribes were uncovered. The so-called 'Lockheed bribery scandal' led to the arrest of Japanese Prime Minister, Kakuei Tanaka, as well as several other officials. The scandal later caused Lockheed board chairman Daniel Haughton, and vice chairman and president Carl Kotchian to resign in February 1976. Tanaka was eventually tried and found guilty of violating foreign exchange control laws, but was not charged with bribery, a more serious criminal offence. Crucially for Lockheed, the fallout from the scandal included the loss of a contract worth in excess of $1 billion.

The Tristar was very similar in proportions to the DC-10; the main difference was the positioning of the third engine, which on the DC-10 was contained totally within the root of the aircraft tail fin, whilst in the

Tristar it was buried in the end of the fuselage. The Tristar's failure to achieve profitability caused Lockheed to withdraw from the civil aircraft business.

After the demise of the De Havilland Comet, US manufacturers produced three- and four-engine jet powered aircraft that dominated intercontinental travel, beginning in 1958 with the introduction of the Boeing 707 and DC-8, powered first by turbojets and later by low bypass ratio turbofans. These were followed in 1970 by the Boeing 747 and in 1972 by the McDonnell Douglas DC-10 and, Lockheed Tristar, and in 1978 by the MD11, all powered by high bypass ratio turbofans. During this period the European aircraft industry was fragmented, with no single manufacturer large enough to take on the might of the American manufacturing giants. This situation changed after the creation of Airbus Industrie in 1970.

Twin-Engine Aircraft

The French, Sud Aviation, Caravelle was the first twin-engine jet powered passenger aircraft to enter service, which it did with Air France in July 1959. The first Caravelles were powered by two Rolls-Royce Avon Mk 522 turbojets (See Fig 50) each delivering 10,500 lbf thrust. The engines were mounted on pylons at the rear of the fuselage. Seating was for 80 passengers, cruising speed was 456 mph at 32,800ft, maximum payload range was 1,600 miles. The engine was later uprated to 11,700 lbf thrust which resulted in an increase in cruising speed to 500 mph and in range

Fig 74: Sud Aviation Caravelle

to 1,750 miles. Several models were produced over the lifetime of the production run as the power of the available engines grew, allowing higher takeoff weights. The largest aircraft was the Caravelle 12 which was powered by Pratt & Whitney JT8-D turbofans of 14,000 lbf thrust. It had accommodation for 140 passengers.

The British Aircraft Association (BAC) introduced a competitor for the Caravelle, the BAC111, which was very similar in design, but powered by Rolls-Royce Spey turbofan engines of 10,400 lbf thrust; this entered service in April 1965. McDonell Douglas also introduced its DC-9 aircraft of similar design in December of that year.

These three aircraft types had the monopoly of jet powered short-haul routes until the emergence of the Boeing 737 in 1968. They were introduced during the period when twin-engine aircraft operation was governed by the FAA 60-minute rule in the USA and the ICAO 90-minute rule elsewhere. The higher cruising speeds of the jet powered twins gave the airline operators more scope in planning routes: 60 minutes to a diversion airport was about 500 miles on a jet powered aircraft, compared with about 380 miles on a turboprop, and 300 miles on a piston powered equivalent.

The Boeing 737-100 was introduced in 1968 as a short to medium range aircraft and went into service with Lufthansa in February of that year. The initial aircraft was powered by JT8D low bypass ratio turbofans of 14,500 lbf thrust as used on the Boeing 727 tri-jet. Accommodation was

Fig 75: Boeing 737-100

for a maximum of 124 passengers, cruising speed was 485 mph and maximum range was 1770 miles. Like the Boeing 707, the engines were mounted under the wings, making the 737 the first twin of this configuration. This was advantageous in that it enabled the incorporation of high bypass ratio engines at a later date allowing the aircraft to grow in capacity and range, a development that was not achievable on the rear engine mounted configurations. The under-wing arrangement later became standard design practice for long-haul twin-engine aircraft.

Both the JT8D-1 and Spey engines developed excellent reputations for reliability. In the first year of operation the JT8D-1 had an IFSDR of 0.16/1,000 hours, and this steadily improved to about 0.03/1,000 hours over the next five years. In the period between 1975 and 1977 the JT8D-15 and JT8D-17 achieved IFSDRs of 0.015/1,000 hours and 0.01/1,000 hours respectively.

Over the first three years of operation in the Trident the Spey achieved an IFSDR of 0.06/1,000 hours. In 1974 the Spey Mk 555 was achieving an IFSDR of 0.03/1,000 hours, which had improved to 0.02/1,000 hours by 1983. Over the same period the JT3D engine achieved an IFSDR of 0.05/1,000 hours in 1965 reducing to 0.025/1,000 hours by 1977.

These levels of reliability were an order of magnitude better than any piston engine had been able to achieve. Surprisingly, in spite of this improvement in reliability, no serious attempt appears to have been made by operators to lobby for an amendment to the 60-minute diversion rule during this period, possibly because the increase in cruise speed of the jet-propelled aircraft effectively doubled the diversion distance relative to when the rule was set in 1953. The FAA and ICAO rules for twin-engine aircraft remained unchanged until 1985.

Table 3 (Appendix 2) shows the comparative performances of a selection of the competing low bypass ratio engines available in the early 1960s

The attractiveness of the Spey relative to the JT8 is apparent, its smaller diameter and length, lower weight and thrust-to-weight ratio and better SFC made it particularly attractive for rear-engine mounted applications. It is therefore not surprising that Boeing were attracted to it for their 727 design; however, in that era other market forces were in play.

The Boeing 737 aircraft has been progressively developed over a period of 50 years and is still in production in 2016. In 1981, Boeing introduced the Classic series (-300, -400 and -500) powered by the high bypass ratio CM 56-3 engine of 20,000 lbf thrust. The 737-400 was a direct competitor for the McDonnell Douglas MD 80, which was a derivative of the DC-9

Fig 76: Boeing 737-800

and introduced in 1980. The Boeing 737 is the best-selling civil aircraft ever; at the end of 2015 about 8,800 had been delivered from a total of 13,250 orders.

The above image of a Boeing 737-800, powered by CFM-56 engines, gives an indication of how much the aircraft has developed compared with the original 737-100 (see Fig 75).

In spite of the success of these low bypass ratio engines, none was able to deliver sufficient thrust to make a viable long range twin-engine aircraft a possibility.

The First Twin-Engine Wide-Body Aircraft

American aircraft produced by Boeing, McDonnel Douglas and Lockheed dominated post WW2 intercontinental air travel for three decades. The fragmented British and European manufacturers were unable to compete. The nationalised British Aircraft Corporation and French company Aerospatiale (previously Sud Aviation) were pre-occupied with the joint venture design and development of the supersonic Concorde, which eventually entered service in 1976 after a protracted and costly design and development programme.

The European aircraft manufacturers realised that to compete with the American industry giants they would have to combine their resources. The result was the formation in 1970 of Airbus Industrie, a consortium

Fig 77: Supersonic Concorde

which was initially composed of Aerospatiale and Deutsche Airbus as 50 percent shareholders, with Hawker Siddeley as a privileged supplier.

By 1970 high bypass ratio engines of in excess of 50,000 lbf thrust were in development for extended range versions of the Boeing 747 and Douglas DC 10, which made the design of a commercially competitive medium range twin-engine powered aircraft a viability. Airbus Industrie took the bold step as its first venture of introducing the world's first wide body, twin-isle, twin-engine aircraft, designated the Airbus A 300.

The first prototype A300 was unveiled on 28 September 1972, making its maiden flight from Toulouse on 28 October that year. The first production model, the A300-B2, entered service in 1974 followed by the A300-B4 one year later. Initially the success of the consortium was poor; by 1979 there were only 81 aircraft in service. The improved comfort offered by the twin-aisle wide-body arrangement had failed to attract passengers in sufficient numbers. Airlines operating the A300 on short-haul routes were forced to reduce frequencies to try and fill the aircraft, and as a result they lost passengers to airlines operating more frequent narrow-body flights.

All this changed with the introduction of airline deregulation in the US. Breaking into the US market was a key goal for Airbus, and was one

Fig 78: Cross section of A300 Fuselage: First Wide-Body Twin

of the main reasons that they had sought the advice of the American Airlines Vice President for Development in the late 1960s when the design was still being formalised. With deregulation, free market forces now determined routes, fares and frequencies on the US East Coast. Eastern Airlines faced severe competition from its competitors National and Delta Airlines on its prized routes between the northeast US and Florida. Eastern was already operating the Lockheed Tristar on the routes and was seeking a 200-seat aircraft with better operating economics. With only two engines, the fuel burn on the A300 was already lower than that of the Tristar, and with two-thirds of the seating capacity Airbus felt they had an ideal product to meet Eastern's needs.

Eastern had already been in discussions with Boeing, Lockheed, and McDonnell Douglas, but its financial position in the wake of the competition with National and Delta meant it needed an aircraft with better economics than the Tristar; it also needed a good commercial deal. Airbus was included in the final sales presentations to Eastern's management and its president, former Apollo 8 mission commander and astronaut Frank Borman. Airbus had four completed A300s with no buyers, and the Airbus team offered to lease these to Eastern for a four-month trial. If they were happy with the aircraft, they could order more. After due consideration Borman sent word to Toulouse: "Congratulations, you've got a blue-eyed baby". It was the breakthrough that saved the A300 programme. After a very successful trial, Eastern ordered 23 A300s on 6 April 1978, with options on more.

The Eastern deal with Airbus was very unpopular in the US. The Department of Commerce and some in the US government demanded an explanation from Borman. A precedent existed where political influence several years earlier had been successful in steering Western Airlines away from the A300. Borman was accused of being anti-patriotic, despite his military and space records. Charles Forsyth, the VP for McDonnell Douglas, is reputed to have contacted Borman directly and accused him of being unpatriotic. Borman reportedly asked Forsyth what brand of luxury car he drove, knowing the answer wasn't American. Borman explained that the most valuable part of the Airbus A300 were its GE engines, along with a large amount of American content from thousands of subcontractors. Congressional hearings were held at Boeing's urging, with accusations of predatory pricing by Airbus to make the sales. None of these manoeuvres succeeded; a major achievement for the young Airbus Industrie. It would become legend throughout Europe how an American astronaut saved the A300 program.

In 1984 Pan Am announced that it would spend more than $1 billion to acquire or lease a total of 28 A300 airliners from Airbus Industrie between 1987 and 1990. These two orders gave Airbus the foothold in the American market that they desired.

During the design of the aircraft it was originally proposed to use the Rolls-Royce RB 207, which evolved into the RB211, thus making it an all European project. However, with lack of foresight, the then state-owned Rolls-Royce withdrew from the project to concentrate on the Tristar. With the exception of a part share in the International Aero Engines (IAE) V2500 on the A320, it would be another 20 years before a Rolls-Royce engine was selected to power an Airbus product. The first A300 aircraft went into service powered by either CF6-50 or Pratt & Whitney JT9D engines, according to customer preference. Accommodation was for up to 345 single-class passengers and operating range was initially 4,140 miles; this was later increased to 4,685 miles on the A300-600R, which had the benefit of increased engine thrust.

Despite its slow start the A300 eventually sold well, reaching a total of 561 delivered aircraft. It was popular with Asian airlines and customers included Korean Air, China Eastern Airlines, China Airlines, Thai

Fig 79: Airbus A300-600

Airways International, Malaysia Airlines, Philippine Airlines, Singapore Airlines, Garuda Indonesia, Pakistan International Airlines, Indian Airlines and Trans Australian Airlines. Asian airlines were operating under the ICAO 90-minute rule as opposed to the FAA 60-minute rule, and this enabled them to use the A300 for routes across the Bay of Bengal and South China Sea.

Airbus followed up the A300 with the A310, a reduced capacity aircraft, with a medium and long range capability. The first aircraft entered service with Swissair and Lufthansa in 1983. The reduced capacity was achieved by reducing the length of the A300 fuselage. Airbus used this basic fuselage diameter for its future derivatives the A330 and A340. In the early years of production, the A300 failed to achieve its full potential due to the imposition of the 60-minute rule, which effectively prohibited transatlantic operations, but this was to change with the eventual introduction of ETOPS.

Chapter 5

Civil Jet Engine Powered Aircraft

ETOPS Era

Whilst Airbus had been attempting to infiltrate the US market with its A300, Boeing had been designing replacements for its ageing 707 and 727 airliners. After looking at alternate configurations Boeing settled on the favoured twin-engine under-wing configuration. The Boeing 707 replacement was the 767-200, powered by two Pratt & Whitney JT9-D engines it entered service with United Airlines on 8 September 1982 on the Chicago to Denver route. The GE CF6-powered 767-200 commenced service three months later with Delta Air Lines.

The 767 was Boeing's first wide-body, twin-engine aircraft and was a direct competitor for the Airbus A300. Later aircraft were fitted with the Rolls-Royce RB211-524G according to customer preference. Capacity and range have been progressively increased from 290 to 375 single-class passengers, and 4,400 to 7,400 miles respectively.

Fig 80: Boeing 767-200

Fig 81: Boeing 757-200

Boeing also introduced the 757, a narrow-body replacement for the Boeing 727.

Like the 767, the 757-200 was a twin-engine under-wing configuration. It had accommodation for 239 single-class passengers and a range of 4,100 miles; the 757-300 had accommodation for up to 295 single-class passengers at a slightly reduced range of 3,900 miles. The aircraft were powered by either the Pratt & Whitney 2037 or Rolls-Royce RB211- 535 according to customer preference. The 757-200, the original version of the aircraft, entered service with Eastern Air Lines in 1983. The first engine to power the 757-200, was the Rolls-Royce RB211-535C, this being the first non-US

Fig 82: RB 211- 535E4

manufactured engine to be selected for the launch of a new Boeing aircraft. This engine was succeeded by the upgraded RB211-535E4 in October 1984; other engines used include the RB211-535E4B, along with the Pratt & Whitney 2037 and 2040.

By 1985 the efficient operation of the Boeing 737, 757 and 767 together with the Airbus A300 and A310, was being compromised by the prevailing FAA and ICAO rules for twin-engine aircraft. Sufficient data on the reliability of jet engines had been accumulated to show a significant reduction in IFSDRs relative to the piston engine data on which the rules had originally been formed.

After the initial teething troubles associated with the introduction of a new engine type, the IFSDR of the high bypass ratio turbofans was proving to be on a par with that of their low bypass ratio predecessors. Unlike the piston engine experience, where IFSDR deteriorated proportionally with increase in horsepower, there appeared to be no deterioration in IFSDR associated with increases in thrust. This is perhaps explained by the fact that increased horsepower was primarily achieved by increasing the number of cylinders and associated components, whereas increases in turbofan thrust were achieved by increasing the bypass ratio, overall pressure ratio, mass flow and the physical size of components rather than the number and complexity. Availability of higher strength nickel and titanium alloys, together with sophisticated turbine cooling methods, enabled significant increases in overall pressure ratio and turbine entry temperature. The emergence of powerful computer programmes facilitated improved aerodynamic and stress analyses, resulting in a beneficial effect on engine and aircraft efficiency.

It was obvious to operators that the full potential of these wide-body twins would only be realised if they could operate on the longer routes, without the constraints of the FAA 60-minute rule and the ICAO 90-minute rule. The engine reliability statistics being achieved indicated that safe flights up to a diversion time of 120 minutes would be achievable, and if so would enable virtually unrestricted transatlantic flights. Recognising these facts, the FAA introduced FAR 121 in 1985 that set out the guide lines for Extended Twin-Engine Operations.

Table 4 (Appendix 2) shows some comparative features and performance of the initial high bypass ratio turbofans and their developed versions that made twin-engine operation achievable. It can be seen from the data that increases in thrust of almost 50 percent were achieved with minimal change to the basic engine configuration. The RB211-524H had no additional rotor stages, whilst the Pratt & Whitney 4000 added an extra LPC Booster stage and the CF6-80 C-2 replaced two HPC stages with three additional LPC Booster stages. The benefit of the 3-shaft arrangement can be appreciated by the fact that the 524H achieved an OPR of 34.5:1 with 14 compression stages including the fan root, whilst the PW4060 achieved an OPR of 31.5:1 in 16 stages and the CF6-80C-2 achieved 30.4:1 in 19 stages. The 524 also had one less turbine stage than the Pratt & Whitney 4060 and two less than the CF6-80C-2. The 524 still had the lowest thrust-to-weight ratio but the difference relative to the competition was reduced relative to the 40,000 lbf thrust engines.

Evolution of the ETOPS Process

Many countries have their own regulating airworthiness authorities, but as far as ETOPS is concerned, we are mainly interested in those that issue the certificates of airworthiness for civil passenger carrying aircraft, and the engines that deliver the power; this narrows the field down to the USA and Europe.

In the US the Federal Aviation Administration (FAA) was formed in 1967, having previously been the Federal Aviation Agency. The FAA is responsible for all issues relating to passenger carrying aircraft in the US, including the granting of a certificate of airworthiness for each aircraft type, and a Type Certificate for each engine model; in both cases, for aircraft and engines that have been designed and manufactured in the US. The FAA also issues certificates of approval for those aircraft and operators that meet the criteria for ETOPS: primarily Boeing, General Electric and Pratt and Whitney.

In Europe the Joint Airworthiness Authority (JAA) was formed in 1970, and performs a similar task for civil aircraft and engines produced in Europe, primarily Airbus and Rolls-Royce.

In addition, the International Civil Aviation Authority (ICAO) was formed in 1947 to ensure uniformity of practice on an international level.

The FAA took the lead when forming the guidelines and regulations for ETOPS, and with one or two exceptions the JAA and ICAO have followed suit.

The introduction of the '100-mile rule' was described earlier: in 1936 the Federal Aviation Authority, an early predecessor of the Federal Aviation Administration, recognising the vulnerability of passenger carrying aircraft to double engine shutdown, issued the rule that declared that a passenger carrying aircraft must not fly on a route to its destination that would take it more than 100 miles from a suitable diversion airport.

The typical cruise speed of a piston engine aeroplane at that time was about 100 miles per hour, hence the rule equated to about 60 minutes flying time.

In 1953, the Federal Aviation Agency, concerned about the deteriorating reliability of piston engines (Fig 35), issued the 60-minute rule, sometimes called the 60-minute diversion period, that restricted twin-engine aircraft to areas of operations that were 60 minutes from an adequate airport, at the one-engine inoperative cruise speed, in still air, at standard atmospheric conditions. The rule was flexible in that it permitted operations up to 75 minutes if special approval was obtained from the FAA. The approval took into consideration the character of the terrain, the kind of operation and the performance of the aircraft.

The purpose of the rule was to reduce the risk of double engine failure to an acceptable level by restricting the flying time to the alternate airport. The cruise speed of a Douglas DC-3 was about 200 mph, so the rule effectively doubled the diversion distance to the alternate airfield that existed under the 1936 rule. However, on certain routes aircraft were forced to fly a dog-leg course to their destination in order to comply with the regulation, and in some cases they were totally excluded from certain routes due to lack of suitable diversion airports en route. The totally excluded area was called the exclusion zone. The following world map shows the approximate effect of the 60-minute rule on twin-engine operations assuming typical jet-propelled aircraft cruise speeds.

In 1953, the ICAO, having themselves reviewed the piston engine failure rates, chose a slightly more lenient approach and permitted a diversion time of 90 minutes for all aircraft. The ICAO 90-minute ruling was adopted by many non-US regulatory authorities, and many non-US airlines operated under it. It is not clear whether either the FAA or ICAO had in mind any numerical probability objectives for double engine failure when setting the diversion times in 1953. The simplest form of assessing the probability of a double engine failure/flight hour on a twin-engine aircraft, is to square the fleet average IFSDR/hour (Ref: Reliability of Systems and Equipment, R G W Cherry). In 1952 the prevailing IFSDR for all piston engine aircraft was assessed to be 0.28/1,000 hours based on a total of 26 million flying hours; however, this included the Douglas DC-3, which had a better than average IFSDR of 0.09/ 1,000 hours based on 7.4 million flying hours experience. The average rate included the exceptionally high IFSDR of the high horsepower engines being used to power 4-engine aircraft, which had no application in 2-engine aircraft at the time; the average without the DC-3 was 0.36/1,000 hours.

60-minute MDT Restriction Limits Route Opportunities

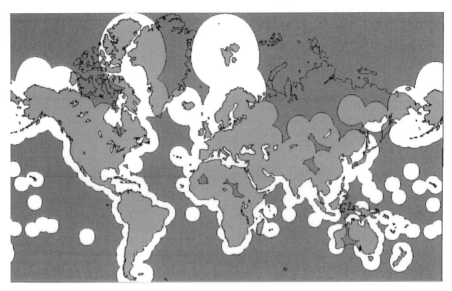

Fig 83: Regions beyond 60-minute threshold (400 nm typical operational range). The white regions indicate the areas that twin-engine planes can travel into, whilst the blue region identifies the exclusion zones where they cannot enter.

Based on this simplistic approach, prior to the introduction of the 60-minute rule, the probability of a double engine failure on a two-hour flight would have been 1.6/10 million hours(1.6×10^{-7}) which was presumably judged acceptable at the time. A similar calculation, using the worst IFSDR of the highest horsepower piston engines of one/1000 hours, gives a probability of double engine failure of 1/million hours (1×10^{-6}), which would be considered an unacceptable level of risk.

The travelling public, whether consciously or not, accepts a level of risk when embarking on a journey in any mode of transport. The safety standards in the design of aircraft, and the engines that power them, are second to none; however, decisions have to be made regarding what is considered to be an acceptable level of risk when assessing potential failure modes that might result in hazardous or catastrophic consequences.

Every effort is made to eliminate such failure modes, but this is not always practicable; for example it is impossible to eliminate all in-flight shut downs (IFSDs), hence there is inevitably some probability of double engine failure, and therefore the objective is to make such an occurrence extremely improbable.

In 1982 the FAA issued Advisory Circular AC25-1309-1 that introduced guidelines for the assessment of potential failure modes and effects of aircraft and their systems. Failure effects were categorized in terms of their severity and numerical probability; these recommendations were updated in 1988 and reissued in AC25-1309-1A.

Failure rates were categorized as follows:

(1) Probable: Probable events may be expected to occur several times during the operational life of each airplane. A probability in the order of one/100,000 hours (1×10^{-5}) or greater is assumed for this category.

(2) Improbable: Improbable events are not expected to occur during the total operational life of a random single airplane of a particular type, but may occur during the total operational life of all airplanes of a particular type. A probability between one/100,000 hours (1×10^{-5}) and one/100 million hours (10^{-8}) is assumed for this category.

(3) Extremely Improbable: Events classified as catastrophic, i.e. resulting in loss of the aircraft, are considered Extremely Improbable and considered so unlikely that they need not be considered to ever occur, unless engineering judgment would require their consideration. A probability in the order of one/billion hours (1×10^{-9}) or less is assumed for this category.

Using the above guidelines from AC 1309-1 and 1A Advisory Circulars, the chart below (Fig 84) summarizes the different failure modes and their probabilities.

By the mid-1980s large bypass ratio turbofans were demonstrating IFSDRs below 0.1/1,000 hours, in particular, the Boeing 757, the Boeing 767 and Airbus A300 wide-body twins were achieving IFSDRs in the region of 0.05/1,000 hours or better. Both aircraft manufacturers and airline operators felt that the existing 60-minute rule, under FAA 121-161, was too restrictive. The rule states that "*Unless authorised by the administrator, based on the character of the terrain, the kind of operation, or the performance of the airplane to be used, no certificate holder may operate two-engine airplanes over a route that contains a point farther than 1-hour flying time (in still air at normal cruising speed with one engine inoperative) from an adequate airport.*"

The manufacturers lobbied for an extension. Boeing was particularly keen to be able to operate the 767 on transatlantic routes without the imposition of the 60-minute diversion time (MDT) that was adversely affecting the payload-range of the aircraft.

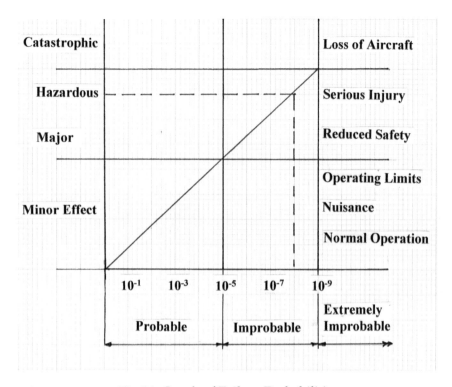

Fig 84: Graph of Failure Probabilities

In addition to the improved engine reliability, the 767 utilised modern technology, computerised systems and a level of redundancy, safety, and efficiency far superior to when the original rule was introduced in 1953. Boeing collected data on the first 767 operations in the first two years of commercial airline services, compiling information on every shutdown and failure of any system, including the engines. In April 1984, Israel Airlines (El Al) became the first airline to operate the 767 on transatlantic services between Montreal and Tel Aviv, but the aircraft's routing complied with the 60-minute rule. Not long after, El Al, Air Canada, and Trans World Airlines (TWA) received exemptions to operate no more than 75 minutes from a suitable diversion airport. This would open up some transatlantic routes and Caribbean routes to the 767. Air Canada was the first to exceed the 60-minute limitation, having earned its 75-minute exemption in 1983.

Fig 85 shows the benefit of the 120-minute rule over the 60-minute rule on a flight from JFK (New York) to LHR (London). The diversion airports are: YYR (Goose Bay, Ontario, Canada); SFJ (Kangerlussuuaq, Greenland) and KEF (Keflavik International Airport, Iceland).

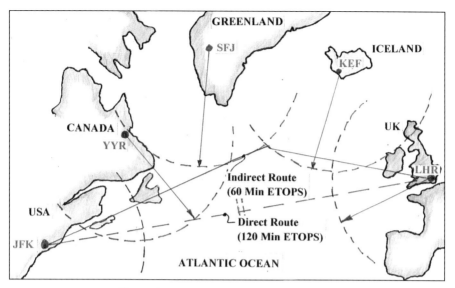

Fig 85: Effect of 120-minute rule on JFK to LHR Flight

In June 1984, Boeing obtained a special concession from the FAA to fly the new 767-200ER (Extended Range) on a 7,500-mile flight from Washington Dulles to Addis Adaba, whilst delivering the aircraft to Ethiopian Airlines. In October 1984, Air Canada took delivery of the first ETOPS qualified 767-200ER, which was permitted to fly 75 minutes from a suitable diversion airport.

The International Civil Aviation Organization (ICAO), the International Federation of Air Line Pilots Association, the US-based pilots' union Air Line Pilots Association (ALPA) and the FAA, made several recommendations to Boeing that resulted in the 767-200ER having a fourth electrical generator independently powered by a hydraulic motor, additional fire suppression features, and equipment for cooling the Cathode Ray Tube (CRT) displays in the cockpit.

By 1985 Boeing was lobbying the FAA for extension of the 75-minute rule to 120 minutes, which would open up a large number of transatlantic routes to the 767. Already several airlines, led by TWA, had petitioned the FAA for an ETOPS extension to 120 minutes, but before the FAA would grant the extension, Boeing had to show "statistical maturity" by equipping a number of 767s with special data gathering equipment to demonstrate unparalleled standards of in-flight reliability; and the Pratt & Whitney JT9-D engines had to log a total of 250,000 consecutive flight hours on passenger flights with a very low rate of shutdown.

In 1985 the FAA agreed to extend the permitted diversion time for twin-engine passenger aircraft to 120 minutes. On 1 February 1985, TWA Flight 810 departed Boston for Paris with eleven observers from the FAA aboard, to complete the first revenue passenger flight in history under the newly issued 120-minute Advisory Circular, FAA-AC120-42.

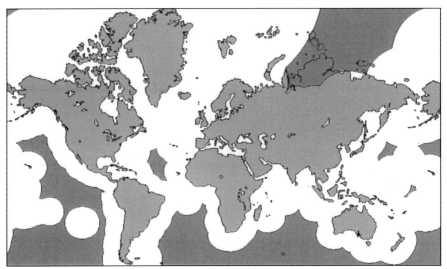

Fig 86: Regions beyond 120-minute ETOPS area (800 nm typical operational range). 120-minute ETOPS Reduced exclusion zones (blue) relative to 60- minute ETOPs (Fig 83)

Before Flight 810 departed, 16 TWA pilots went through specialized ETOPS training on international requirements, intensive time in a simulator, and landing procedures for the airport at Sondrestromfjord in Greenland, the designated 120-minute diversion airport. The fuel burn was found to be 7,000 lb an hour less than that of an L-1011 Tristar on the same route.

Boeing continued to obtain data from its 120-minute ETOPS transatlantic flights with a view to obtaining a further extension to 180-minute ETOPS, which would enable it to fly a direct route from Hawaii to California.

As a consequence, the FAA replaced AC 120-42 with AC 120-42A that extended the permissible ETOPS diversion time for suitably equipped twin-engine aircraft of proven reliability.

The principal factor in extending the diversion time was the continuing reduction of IFSDRs as the high bypass ratio turbojets matured.

180-minute MDT Further Expands Route Opportunities

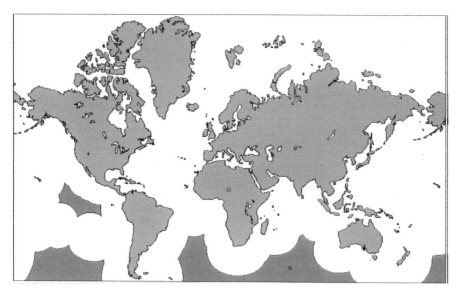

Fig 87: Regions beyond 180-minute ETOPS area
(1200 nm typical operational range)

Obviously a double engine failure is most likely to result in an aircraft accident incurring loss of life, and is therefore classified as a catastrophic event requiring a probability of failure of 1×10^{-9}/flight hour (see above).

The FAA 120-42 method of assessing risk due to double engine failure was by use of a formula generated by the ICAO in 1984, (Ref AN-WP/5593), which assumed the maximum IFSDR should not result in a double engine failure probability worse than the prevailing world civil transport aircraft accident rate over a several year period. The formula used was: -

IFSDR = $\sqrt{10^{-8}}$ (0.6+0.4T)/Ty, where T= intended duration of flight, and y = diversion time.

As an example, a flight of seven hours with a diversion time of two hours indicates a requirement for an IFSDR of 0.05/1,000 (5×10^{-5}) hours. This rate was the level set to qualify for 120-minute ETOPS. When considering what level to apply for 180-minutes ETOPS, the FAA proposed using a risk factor which, when applied to 180-minutes ETOPS, resulted in a maximum IFSDR of 0.02/1,000 hours (2×10^{-5}).

The origin of this formula is lost in time, however in 2001 the JAA (P-NPA-20) proposed a more sophisticated approach to calculating the probability of double engine failure. It recognised that IFSDRs were approximately half the overall rate during cruise, and double the overall rate at maximum continuous power rating. The resulting formula to derive the probability of double engine failure per flight hour for a twin-engine aeroplane is considered more representative of how the remaining engine would have to perform during the diversion period. The resulting formula is:

p/flight hour= [2(Cr x{T-t}) x Mr(t)] ÷T, where p is the probability of a dual independent propulsion unit failure per flight hour on a twin; 2 is the number of opportunities for an engine failure, i.e. 2 engines; Cr is the cruise IFSDR = 0.5 x overall rate; Mr is the max continuous IFSDR = 2 x the overall rate; T is the planned maximum flight duration in hours to the destination airport, and t is the planned diversion time to the identified suitable alternate airport.

The assumptions regarding the cruise and max continuous IFSDRs was based on engine manufacturers historical data for high bypass ratio turbofans, presented to the JAA and ARAC ETOPS / LROPS working groups over a ten-year period up to 2001. Using the above formula, Fig 88 shows the probability of double engine failure versus IFSDR for

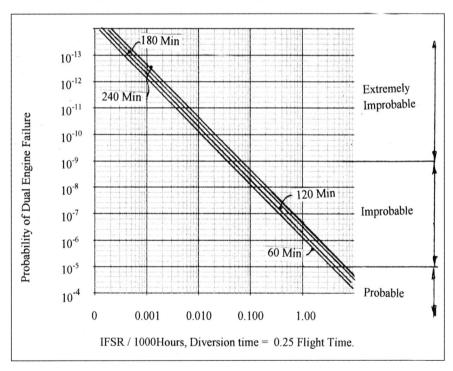

Fig 88: Dual Engine Failure Probability Versus Diversion Time

different diversion times where T=4t. Where T=4t the above formula becomes P=1.5t x IFSDR²

From the graph (Fig 88) it can be observed that the IFSDR of 0.05/1,000 hours recommended in AC120-42A for 120-minute ETOPS does not quite meet an extremely improbable likelihood, having a probability of 7.5×10^{-9} per hour, whilst the rate of 0.02/1,000 hours for 180-minute ETOPS almost satisfies the requirement at 1.8×10^{-9}.

120 and 180-minutes Regulation AC120-42A

The FAA issued AC 120-42A in December 1988, which opened the door for the introduction of 180-minute diversion times. In addition to defining the required propulsion system IFSDRs for 120-minutes and 180-minutes ETOPS, the Advisory Circular (AC) defined the procedures for design, maintenance, training and operation procedures required for the aircraft and engine configuration under consideration; these are summarized below.

The requirements for the propulsion system were stated as follows: "A review of the historical data (1978 through 1988) for transport aviation two-engine turbofan powered large commercial airplanes indicates that the current safety record, as exemplified by the world accident rate (airworthiness causes) is sustained in part by a propulsion system IFSDR of only about 0.02/1000 hours.

Although the quality of this safety record is not wholly attributable to the IFSDR, it is believed that maintaining an IFSDR of that order is necessary to not adversely impact the world accident rate from airworthiness causes.

Upon further review of the historical data base, and in consideration of the required safety of extended range operation, it is necessary that the achieved performance and reliability of the airplane should be shown to be sufficiently high. When considering the impact of increasing diversion time, it must be shown that the operation can be conducted at a level of reliability resulting in no adverse change in risk."

In addition to achieving an approved IFSDR of the engines, the AC introduced other requirements to ensure the safety of the aircraft and engine combination during the extended diversion times as follows:

Design

The design must comply with the FAA fail-safe design concept to ensure that major failure conditions are improbable, and that catastrophic failure conditions are extremely improbable: "In any system or subsystem, the failure of any single element, component, or connection during any one flight should be assumed, regardless of its probability. Such single failures should not prevent continued safe flight and landing, or significantly reduce the capability of the airplane or the ability of the crew to cope with the resulting failure conditions. Subsequent failures during the same flight, whether detected or latent, and combinations thereof, should also be assumed, unless their joint probability with the first failure is shown to be extremely improbable (1×10^{-9}). These principles should include the following:

- designed integrity and quality, including life limits to ensure intended function and prevent failures;
- redundancy or backup systems to enable continued function after any single or other number of failure(s), e.g. provision of two or more hydraulic systems, flight control systems, etc.
- isolation of systems, components, and elements so that the failure of one does not cause the failure of another, isolation is also termed independence;
- proven reliability so that multiple independent failures are unlikely to occur during the same flight;
- failure warning or indication to provide detection;
- flight crew procedures for use after failure detection, to enable continued safe flight and landing by specifying crew corrective action;
- the capability to check the condition of components; designed failure effect Limits, including the capability to sustain damage to limit the safety impact or effects of a failure;
- designed failure path to control and direct the effects of a failure in a way that limits its safety impact;
- margins or factors of safety to allow for any undefined or unforeseeable adverse conditions;
- error tolerance that considers adverse effects of foreseeable errors during the airplane's design, manufacture, operation, and maintenance test.

Alternate Airports

Any airport designated as an en route alternate must have the capabilities, services, and facilities to safely support that particular airplane; the weather conditions at the time of arrival must give a high assurance that adequate visual references will be available upon arrival at decision height

(DH) or minimum descent altitude (MDA), and that the surface wind conditions and corresponding runway surface conditions will be within acceptable limits to permit the approach and landing to be safely completed with an engine and/or systems inoperative.

For an airport to be considered suitable it should have the capabilities, services, and facilities necessary to designate it as an adequate airport, and have weather and field conditions at the time of the particular operation which provide a high assurance that an approach and landing can be safely completed with an engine and/or systems inoperative in the event that a diversion to the en route alternate becomes necessary.

Due to the natural variability of weather conditions with time, as well as the need to determine the suitability of a particular en route airport prior to departure, the en route alternate weather minima for dispatch purposes are generally higher than the weather minima necessary to initiate an instrument approach. This is necessary to assure that the instrument approach can be conducted safely if the flight has to divert to the alternate airport.

Additionally, since the visual reference necessary to safely complete an approach and landing is determined, among other things, by the accuracy with which the airplane can be controlled along the approach path by reference to instruments and the accuracy of the ground-based instrument aids, as well as the tasks the pilot is required to accomplish to maneuver the airplane so as to complete the landing, the weather minima for non-precision approaches are generally higher than for precision approaches. It is a requirement that the suitability of the weather conditions at the proposed alternate airports is monitored during flight.

Maintenance

A special maintenance programme is required aimed at improving reliability and reducing the risk of common cause events, this is summarized below.
- ETOPS maintenance requirements will be expressed in, and approved as, supplemental requirements; this includes maintenance procedures to preclude identical action being applied to multiple similar elements in any ETOPS critical system (e.g. fuel control change on both engines) i.e. multiple independent failures that may occur in the same flight.
- ETOPS related tasks should be identified on the operator's routine work forms and related instructions.
- ETOPS related procedures, such as involvement of centralised mainte-

nance control, should be clearly defined in the operator's program.
- An ETOPS service check should be developed to verify that the status of the airplane and certain critical items are acceptable. This check should be accomplished and signed off by an ETOPS qualified maintenance person immediately prior to an ETOPS Flight.
- Log books should be reviewed and documented as appropriate to ensure proper maintenance procedures, and system verification procedures have been properly performed.

The operator should develop an ETOPS Manual that references the maintenance programs and other requirements described by the advisory circular, and clearly indicate where they are located in the operator's manual system. All ETOPS requirements, including supportive programs, procedures, duties, and responsibilities, should be identified and subject to revision control. This manual should be submitted to the certificate-holding office 60 days before implementation of the ETOPS flights.

The maintenance programme must include the following:
- A means of monitoring engine and Auxiliary Power Unit (APU) oil consumption.
- An engine condition monitoring system which records engine operating parameters that will detect engine deterioration and enable engine replacement before engine operation limits are reached. (e.g. leaving an engine on wing until it is rejected by reaching its red-line limit of shaft speed or temperature would not be permissible in an ETOPS operation).
- The operator should develop a verification program, or procedures should be established to ensure corrective action following an engine shutdown, primary system failure, adverse trends or any prescribed events which require verification flight or other action and establish means to assure their accomplishment, a clear description of who must initiate verification actions, and the section or group responsible for the determination of what action is necessary should be identified in the program. Primary systems, like the APU, or conditions requiring verification actions should be described in the operators ETOPS maintenance manual.

Reliability

An ETOPS reliability program should be designed with early identification and prevention of ETOPS related problems as the primary goal. The program should be event orientated and incorporate reporting procedures for significant events detrimental to ETOPS flights. This information should be readily available for use by the operator and FAA to help establish that the reliability level is adequate, and to assess the operator's competence and capability to safely continue ETOPS. The FAA certificate-holding district office should be notified within 72 hours of events reportable through this program which should be included in the following items: in-flight shutdowns; diversion or turn back; uncommanded power changes or surges; inability to control the engine or obtain desired power; problems with systems critical to ETOPS; any other event detrimental to ETOPS; report should identify the following; airplane identification (type and N-Number); engine identification (make and serial number); total time, cycles, and time since last shop visit; for systems, time since overhaul or last inspection of the discrepant unit; phase of flight; corrective action.

Propulsion System Monitoring

A process is required to identify the corrective actions necessary in the event that the propulsion system IFSDR (computed on a 12-month rolling average) exceeds 0.05/1,000 engine hours for a 120-minute operation, or exceeds 0.02/1,000 engine hours for a 180-minute operation. An immediate evaluation and recovery plan should be agreed with the certification authority.

Training

A special ETOPS oriented training plan should be introduced.

Parts Control

A parts control programme must be introduced in order to ensure that only ETOPS approved parts are fitted to ETOPS approved aircraft.

Operation Approval

The following criteria are required for operational approval of extended range operations with a maximum diversion time of 120 minutes to an en route alternate airport (at single-engine inoperative cruise speed in still air).

Requesting Approval

It is required that any operator seeking approval for 120-minute ETOPS should submit the requests, with the required supporting data, to the certificate-holding district office at least 60 days prior to the proposed start of extended range operation with the specific airframe/engine combination. In considering an application from an operator to conduct extended range operations, an assessment should be made of the operator's overall safety record, past performance, flight crew training, and maintenance programs. The data provided with the request should substantiate the operator's ability and competence to safely conduct and support these operations. Any reliability assessment obtained, either through analysis or service experience, should be used as guidance in support of operational judgments regarding the suitability of the intended operation.

Assessment of the Operator's Propulsion System Reliability

Following the accumulation of adequate operating experience by the world fleet of the specified airframe-engine combination, and the establishment of an IFSDR objective of 0.05/1,000 hours necessary for 120-minute extended range operations, an assessment should be made of the applicant's ability to achieve and maintain this level of propulsion system reliability. This assessment should include trend comparisons of the operator's data with other operators as well as the world fleet average values, and the application of a qualitative judgment that considers all of the relevant factors. The operator's past record of propulsion system reliability with related types of power units should also be reviewed, as well as its record of achieved systems reliability with the airframe/engine combination for which authorisation is sought to conduct extended range operations.

Engineering Modifications and Maintenance Program

The following items, as part of the operator's program, will be reviewed to ensure that they are adequate for extended range operations: a list of all modifications included specifically to gain ETOPS approval must be provided to the regulating authority; the approved changes to the maintenance and training procedures, practices, or limitations established to qualify for extended range operations should be submitted 60 days before such changes may be adopted.

Reliability Reporting

The reliability reporting program as supplemented and approved, should be implemented prior to and continued after approval of extended range operation. Data from this process should result in a suitable summary of problem events, reliability trends and corrective actions and be provided regularly to the certificate holding district office (see above); a procedures and centralised control process should be established which would preclude an airplane being dispatched for extended range operation after propulsion system shutdown or primary airframe system failure on a previous flight, or significant adverse trends in system performance, without appropriate corrective action having been taken. Confirmation of such action as being appropriate, in some cases, may require the successful completion of one or more non-revenue or non-ETOPS revenue flights (as appropriate) prior to dispatch on an extended range operation; the program used to ensure that the airborne equipment will continue to be maintained at the level of performance and reliability necessary for extended range operations; the engine condition monitoring program and the engine oil consumption monitoring program.

Flight Dispatch Considerations

The unique nature of extended range operations with two-engine airplanes necessitates a re-examination of the flight dispatch operations to ensure that the approved programs are adequate for the purpose as follows: System redundancy levels appropriate to extended range operations should be included in the Master Minimum Equipment List and are considered dispatch critical for ETOPS operation, these include electrical (including) battery; hydraulics; pneumatics flight instrumentation; fuel; flight control; ice protection; engine start and ignition; propulsion system instruments; navigation and communications; auxiliary power unit (APU); air conditioning and pressurisation; cargo fire suppression; emergency equipment and any other equipment necessary for ETOPS.

Communication and Navigation Facilities

An airplane should not be dispatched on an extended range operation unless: communications facilities are available to provide under normal conditions of propagation, at the normal one engine inoperative cruise altitudes (typically 10,000ft), reliable two-way voice communications between the airplane and the appropriate air traffic control unit over the planned route of flight, and the routes to any suitable alternate to be used in the event of diversion; non-visual ground navigation aids are available and located so as to provide, taking account of the navigation equipment installed in the airplane, the navigation accuracy necessary for the planned route and altitude of flight, and the routes to any alternate and altitudes to be used in the event of an engine shutdown; and visual and non-visual aids are available at the specified alternates for the authorised types of approaches and operation minima.

Fuel and Oil Supply

An airplane should not be dispatched on an extended range operation unless it carries sufficient fuel and oil to meet the requirements of FAR Part 121, and any additional fuel that may necessary to complete the flight under any emergency ETOPS conditions. In computing fuel requirements, advantage may be taken of drift-down, and at least the following should be considered as applicable:- Current forecast winds and meteorological conditions along the expected flight path at one engine inoperative cruising altitude and throughout the approach and landing; any necessary operation of ice protection systems and performance loss due to ice accretion on the unprotected surfaces of the airplane; any necessary operation of the auxiliary power unit (APU); loss of airplane pressurisation and air conditioning; consideration should be given to flying at an altitude meeting oxygen requirements in the event of loss of pressurisation; an approach followed by a missed approach and a subsequent approach and landing; the navigational accuracy necessary: and any known Air Traffic Control (ATC) constraints.

Critical Fuel Reserves

In establishing the critical fuel reserves, the applicant is to determine the fuel necessary to fly to the most critical point and execute a diversion to a suitable alternate airport under the conditions outlined below.

If it is determined that the fuel to complete the critical fuel scenario exceeds that for a normal flight as defined in FAR Part 121, additional fuel should be included to the extent necessary to safely complete the critical fuel scenario which should allow for a contingency figure of five percent added to the calculated fuel burn from the critical point to allow for errors in wind forecasts, a five percent penalty in fuel mileage, any Configuration Deviation List items, both airframe and engine anti-icing; and account for ice accumulation on unprotected surfaces if icing conditions are likely to be encountered during the diversion. If the APU is a required power source, then its fuel consumption should be accounted for during the appropriate phase(s) of flight.

The Critical Fuel Scenario is described as follows: At the critical point, consider simultaneous failure of an engine and the pressurisation system (critical point based on time to a suitable alternate airport at the approved one-engine inoperative cruise speed); immediate descent to and continued cruise at 10,000ft at the approved one-engine inoperative cruise speed, or continued cruise above 10,000ft if the airplane is equipped with sufficient supplemental oxygen in accordance with FAR Section 121.329; upon approaching destination, descent to 1,500ft above destination, hold for 15 minutes, initiation of an approach followed by a missed approach and then execution of a normal approach and landing.

Alternate Airports

An airplane should not be dispatched on an extended range operation unless the required take off, destination and alternate airports, including suitable en route alternate airports to be used in the event of engine shutdown or airplane system failure(s) which require a diversion, are listed in the cockpit documentation (e.g., computerised flight plan). Suitable en route alternates should also be identified and listed in the dispatch release for all cases where the planned route of flight contains a point more than one-hour flying time at the one-engine inoperative speed from an adequate airport.

Since these suitable en route alternates serve a different purpose than the destination alternate airport, and would normally be used only in the event of an engine failure or the loss of primary airplane systems, an airport should not be listed as a suitable en route unless:

- The landing distances required as specified in the Aircraft Flight Manuel (AFM) for the altitude of the airport, for the runway expected to be used, taking into account wind conditions, runway surface conditions, and airplane handling characteristics, permit the airplane to be stopped within the landing distance available as declared by the airport authorities.
- The airport services and facilities are adequate for the applicant operator's approved approach procedure(s) and operating minima for the runway expected to be used, and the latest available forecast weather conditions for a period commencing one hour before the established earliest time of landing and ending one hour after the established latest time of landing at that airport, equals or exceeds the authorised weather minima for en route alternate airports.
- In addition, for the period commencing one hour before the established earliest time of landing, and ending one hour after the estab-

lished latest time of landing at that airport, the forecast crosswind component, including gusts, for the landing runway expected to be used, should be less than the maximum permitted crosswind for landing; during the course of the flight, the flight crew should be informed of any significant changes in conditions at designated en route alternates.
- Prior to a 120-minute extended range flight proceeding beyond the extended range entry point, the forecast weather for the above mentioned time periods, landing distances, and airport services and facilities at designated en route alternates should be evaluated. If any conditions are identified (such as weather forecast below landing minima) which would preclude safe approach and landing, then the pilot should be notified and an acceptable alternate(s) selected where safe approach and landing can be made.

Airplane Performance Data

No airplane should be dispatched on an extended range flight unless the operator's Operations Manual contains sufficient data to support the critical fuel reserve and area of operations calculation. The following data should be based on FAA-approved information provided or referenced in the Airplane Flight Manual:
- Detailed one-engine inoperative performance data, including fuel flow for standard and non-standard atmospheric conditions and as a function of airspeed and power setting, where appropriate, covering, drift down (includes net performance); cruise altitude coverage including, 10,000ft; holding; altitude capability (includes net performance) and missed approach.
- Detailed all engine operating performance data, including nominal fuel flow data, for standard and non-standard atmospheric conditions and as a function of airspeed and power setting, where appropriate, covering: - Cruise (altitude coverage including 10,000ft) and holding.
- Details of any other conditions relevant to extended range operations which can cause significant deterioration of performance, such as ice accumulation on the unprotected surfaces of the airplane, RAM Air Turbine (RAT) deployment, thrust reverser deployment, etc.; the altitudes, airspeeds, thrust settings, and fuel flow used in establishing the ETOPS area of operations for each airframe-engine combination must be used in showing the corresponding terrain and obstruction clearances in accordance with FAR Section 121.191.

Flight Crew Training, Evaluation, and Operating Manuals

The FAA will review in service experience of critical and essential airplane systems. The review will include system reliability levels and individual event circumstances, including crew actions taken in response to equipment failures or unavailability's. The purpose of the review will be to verify the adequacy of information provided in training programs and operating manuals. The aviation industry should provide information for and participate in these reviews. The FAA will use the information resulting from these reviews to modify or update flight crew training programs, operating manuals, and checklists, as necessary.

The operator's training program in respect to extended range operations should provide training for flight crew members followed by subsequent evaluations and proficiency checks in the following areas:

- Performance: flight planning, including all contingencies and flight performance progress monitoring.
- Procedures: diversion procedures; use of appropriate navigation and communication systems; abnormal and emergency procedures to be followed in the event of foreseeable failures, including procedures for single and multiple failures in flight that would precipitate go/no-go and diversion decisions; operational restrictions associated with these failures including any applicable Maintenance Equipment List (MEL) considerations; procedures for air start of the propulsion systems, including the APU, if required; crew incapacitation; use of emergency equipment including protective breathing.
- Conditions at designated en route alternates which would preclude safe approach and landing.
- Understanding and effective use of approved additional or modified equipment required for extended range operations.
- Fuel management procedures to be followed during the en route portion of the flight; these procedures should provide for an independent cross check of fuel quantity indicators, for example, fuel flows could be used to calculate fuel burned and compared to the indicated fuel remaining.

ETOPS Check Airman

The operator should designate a specific ETOPS Check Airman whose objective is to ensure standardized flight crew practices and procedures, and also to emphasise the special nature of ETOPS operations. Only airmen with a demonstrated understanding of the unique requirements of ETOPS should be designated as a check airman.

180-Minutes ETOPS Special Conditions

In addition to the above requirements the following additional conditions apply for operations requiring 180-minutes diversion time:

- Each operator requesting approval to conduct extended range operations beyond 120 minutes should have approximately 12 consecutive months of operational in-service experience with the specified ETOPS configured airframe-engine combination in the conduct of 120-minute operations. The substitution of in-service experience which is equivalent to the actual conduct of 120-minute operators will be established by the Director, Flight Standards Service, on a case by case basis.
- Prior to approval, the operator's capability to conduct operations and implement effective ETOPS programs in accordance with the criteria detailed in Paragraph 10 of the advisory circular will be examined. Only operators who have demonstrated capability to conduct a 120-minute program successfully will be considered for approval beyond 120 minutes. These operators should also demonstrate additional capabilities discussed in this paragraph.
- Approval will be given on a case by case basis for an increase to their area of operation beyond 120 minutes. The area of operation will be defined by a maximum diversion time of 180 minutes to an adequate airport at approved one engine inoperative cruise speed (under standard conditions in still air). The dispatch limitation will be a maximum diversion time of 180 minutes to a suitable airport at approved single-engine inoperative speed (under standard conditions in still air).

The ETOPS propulsion system approval considerations are as summarised below.

(1) Up to 120-Minute Operation:
- 250,000 engine hours (significant portion with experience candidate airplane).
- Achieve an IFSDR of approximately 0.05/1,000 hours, (the objective is continuing towards a rate of 0.02/1,000 hours).
- A periodic review of propulsion system data and service experience.

(2) Greater than 120-Minute Operation:
- 250,000 engine hours (significant portion with experience candi-

date airplane) plus an additional 12 months with ETOPS operation on candidate aeroplane.
- Achieve and maintain an IFSDR of approximately 0.02/1000 hours.
- A periodic review of propulsion system data and service experience.

In March 1989 the FAA approved the GE-CF6 powered Boeing 767 ER as the first aircraft to operate under the new 180-minute ETOPS rules. Approvals were progressively applied to the other engine and aircraft combinations to the extent that twin-engine operations became extensively used on Transatlantic and Pacific routes.

In 1993 the FAA reviewed the comparative operation of a sample of 2, 3 and 4-engine aircraft with respect to engine reliability following the introduction of AC120-42. (Ref AD-A274-860, Aircraft Engine Reliability and Inspection Investigations). Information was collected between February 1988 and January 1991 from several operators of the 4-engine Boeing 747, the 3-engine Douglas DC-10 and the 2-engine Airbus A310 and Boeing 767. The achieved average IFSDR/1000 hours are summarised as follows:

JT9-D Powered:
    Boeing 747 = 0.152
    DC-10 = 0.26
    Boeing 767 = 0.062

CF6 Powered:
    Boeing 767 = 0.013
    DC-10 = 0.064
    A300 = 0.054

Analysis of the data reveals that the 2-engine aircraft had significantly better IFSDRs than either of the 3- and 4-engine aircraft. The CF6 powered Boeing 767 IFSDR met the 180-minute ETOPS requirement whilst the JT9-D powered 767 and CF6 powered A300 just failed to meet the 120-minute ETOPS requirement. The A300 CF6 progressively improved from 0.08 to 0.03 between 1988 and 1991. This is early evidence of the beneficial effects of ETOPS procedures on engine reliability. The introduction of ETOPS procedures increased the cost of operation, and airline operators were reluctant to read these across to their 3- and 4-engine aircraft. Also modifications to address known failures that resulted in IFSD had a higher incorporation priority on the ETOPS aircraft than on 3- and 4-engine aircraft; however, the latter ultimately benefited from the ETOPS process.

A Boeing article entitled-ETOPS Maintenance on Non-ETOPS Airplanes- presents data from United Airlines operations to show how engine reliability of its non-ETOPS Boeing 747 fleet was improved by adopting maintenance procedures and practices from its ETOPS approved Boeing 767 fleet. During a ten-year period from 1989-99, the IFSDR of the Pratt & Whitney JT9-D engines on the Boeing 747-100 aircraft was reduced from 0.062/1,000 hours to 0.032/1,000 hours.

The ICAO, JAA and other regulating bodies used the rules outlined in the FAA Advisory Circular to formulate their own Regulations. Over the next decade the Boeing 737, 757 and 767 series and the Airbus A300, A310, A320 and A330 series were approved for ETOPS operations, and as a consequence the 3-engine Douglas DC10 and MD11 together with the Lockeed L1011 became obsolete.

Chapter 6

Early-ETOPS and Beyond

In March 1993 Airbus Industrie launched the A340-200 into service with Lufthansa. The Aircraft was designed for long-haul operation over a range of 7,700 miles. With a maximum takeoff weight of 600,000 lb there was no engine available of high enough thrust to enable it to be configured as a twin; consequently, the initial aircraft were powered by four CFM 56 engines each rated at 34,000 lbf thrust.

As early as 1986 Boeing had commenced project designs for a larger, longer range version of the 767 which it designated the 767 X, the objective being to provide a twin-engine aircraft to replace the ageing early versions of the Boeing 747. Additionally, it was intended as a direct competitor of the Airbus A340. The initial reaction from airlines was less than enthusiastic and the outcome was the formation of a consortium of eight major airlines from different parts of the world: All Nippon Airways, American Airlines, British Airways, Cathay Pacific Airways, Delta Airlines, Japan Airlines, Qantas and United, all of whom participated in the design phase and had a role in the development of the airliner.

A Working Together Group was set up involving these airlines and representatives from Boeing and the engine companies Rolls-Royce, Pratt & Whitney and General Electric. The Boeing 777 twin-engine aircraft evolved from their deliberations. One of the key customer requirements was that the aircraft should have 180-ETOPS approval at aircraft certification in order that it could fly the routes that the replaced 4-engine Boeing 747s were currently operating. The requirement was dubbed "ETOPS Out of the Box", and is now referred to as Early-ETOPS. In order to achieve the required IFSDR of 0.02/1,000 hours for 180-minutes ETOPS, it was fundamental that the higher thrust engines should be a derivative of an engine of proven reliability.

After discussion between Boeing, the engine manufacturers, airline customers and the FAA, a set of special conditions was agreed aimed at ensuring that engine IFSDRs would not be compromised relative to the previously agreed process based upon service experience. The special Conditions issued by the FAA, under 25 ANM-84, included requirements for engine and flight testing aimed at confirming the suitability of the engine and aircraft combination for 180-minutes ETOPS at entry into service. Boeing agreed a test procedure aimed at ensuring that engine reliability would meet the required standard. As part of the engine certification process the engine manufacturer produces a Failure Mode and Effect Criticality Analysis (FMECA), whose aim is to identify the failure modes that might affect the safety of the aircraft. Historically, catastrophic failure modes would have to meet the extremely improbable likelihood (1×10^{-9}). In some instances, a failsafe design would be adopted to meet this objective, for example loss of an engine due to failure of an engine mount would be prevented by incorporation of a secondary load path. Hazardous failure modes, e.g. the uncontained failure of a rotor disc, would have to meet an extremely remote (1×10^{-8}) probability; to satisfy this requirement, rotor discs are classified as critical components and subjected to rigorous inspection and test techniques, and are life limited in service. Parts or systems whose failure might cause an IFSD would be expected to meet a probability of failure of no worse than Probable (1×10^{-6}). Dedicated bench and engine testing would be relied upon to expose any discrepancies in the probability assumptions.

As part of the Early-ETOPS design process it was a requirement to review all known IFSD events over a period of the previous ten years on those engines in service from which the candidate engine was derived, and to ensure that where applicable, solutions to these problems were included in the engine specification. Identified failure modes that might give rise to IFSDs were to be given particular attention, being either eliminated by redesign, or if this were impractical, the assumptions regarding their probability of failure being validated during the test programme. The use of new technology in the design had to be backed up by sufficient testing to expose any problems that might result in IFSD events. At the time, the progressive use of increasingly sophisticated computer programmes and greater computer processing capability, had a positive effect on prediction methods; e.g. finite element stress analyses, heat transfer predictions, vibration analyses and aerodynamics, all of which enabled a more accurate assessment of possible failure modes and problem resolution.

The engine development programme was extended to increase the number of engine test cycles performed prior to EIS, and the cycle content was amended to make it more representative of the in-service environment, the aim being to eliminate the type of problems that typically occur early after the introduction of a new engine type. Prior to engine certification a dedicated engine had to complete 3,000 engine cycles with the production standard of engine components. Modifications to the production standard of components post the completion of this test could only be incorporated under exceptional circumstances, this prohibited the 'shoe-horning' in of non-essential modifications without adequate cyclic testing, which had been the practice hitherto. Unlike previous engine cyclic tests in engine certification programmes, the test had to be carried out with the complete propulsion system, including an operable thrust reverser and all the nacelle and engine build units, including bleed air ducting, nacelle anti-icing, nacelle ventilation etc.

The start-stop cycles included a mix of maximum take-off, de-rated take-off and maximum continuous thrust levels that represented typical operations expected in the first two years of service for the specific model, recognising the proposed route structures of the likely first customers, with various missions being uniformly distributed through the test. Each cycle included ground idle, taxi, take-off, climb, cruise, descent (minimum idle), approach (approach idle and glide slope intercept) and landing, including reverse thrust. Unlike previous cycle testing, each cycle included engine start and shutdown, during which the main rotor speeds should not be greater than that due to windmilling rotation resulting from prevailing ambient wind conditions. A minimum of 50 of these starts were to occur after the engine had been shut down for a period of a minimum of three hours to make them more representative of the likely in service scenario. Typical cabin and anti-icing bleeds were taken throughout the test, along with power extraction for hydraulics and electrics etc.; approximately one-third of the 3,000 test cycles were performed with the engine anti-ice bleeds active. Three simulated diversions of 180 minutes were included, being equally spaced throughout the test, the final one being within 100 cycles of the end of the test. Each diversion cycle had to include the air and power extractions mentioned above. The maximum continuous rating chosen had to be the highest envisaged during a 180-minutes ETOPS diversion.

In order to ensure that the test replicated the possible in-service rotor induced vibration levels, the main engine rotors had weights added to induce out of balance forces throughout the test that would induce vibratory excitation greater than 90 percent of the recommended in-service maintenance vibration limits; this had the effect of exposing any vulnerable external components to greater levels of excitation than they would have typically experienced on a cyclic endurance test, thereby exposing any potential weaknesses.

The cyclic test also included vibration endurance speed steps of 60 rpm increments of the high speed rotor's operating range for the equivalent of three million (3×10^6) cycles for steady-state rotor speed points that represent take-off, climb, cruise, descent, approach, landing and thrust reverse; and 300,000 (3×10^5) cycles for rotor operational speed points in the range between flight idle and cruise. The objective of this part of the test was to expose the rotor blades and stator vanes to any inherent individual resonances that occur in the running rage to at least 1×10^7 endurance cycles at their resonant frequency. Routine maintenance operations were carried out during the test to the procedures laid down in the official maintenance manual.

On completion of the cyclic test the engine had to be inspected for comparison with the in-service on-wing inspection limits; following this the propulsion system was completely stripped and laid out for inspection, and the condition of parts recorded and correlated with the proposed in-service on-condition monitoring acceptance limits.

Hardware conditions that could result in loss of thrust control, in-flight shutdown or other power loss within a period of operation before the component, assembly, or system would likely have been inspected, or functionally tested for integrity while in service, were not acceptable. To limit the risk of this eventuality jeopardising the successful completion of the test, which was by necessity late in the programme, the definitive test was preceded by other cyclic tests aimed at exposing any potential problems in time for recovery action to be taken. By the time that the competing engines entered service the test engines had completed more cycles than any other previous engine type.

An ex-development cyclic engine was installed on a dedicated aircraft within the flight certification test programme, which was designated to complete a further 1,000 flight cycles; the aim was to expose the engine to cyclic conditions which could not be readily repeated during the bench testing. The certification testing included simulated 180-minute diversions with one engine inoperative, this enabled finalisation of the engine operating parameters and operating procedures in the event of an in-service diversion being required.

In order to prove the acceptability of post certification modifications a 3,000 cycle engine was included in the ongoing engine test programme. In the event that the Auxiliary Power Unit (APU) was an essential part of the ETOPS certification, then it was a requirement that it also should be subjected to a similar 3,000 cycle test.

United Airlines was the launch customer for the Boeing 777-200. They received their first aircraft on 15 May1995, and the FAA awarded 180-minute ETOPS for the Pratt & Whitney engine powered aircraft on 30 May 1995, making it the first airliner to carry an ETOPS-180 rating at EIS. This was followed by British Airways with GE 90-77B engines in November 1995 and Thai Airways with the RR Trent 877 in March 1996. The Boeing 777 proved to be popular, to the detriment of the Airbus A340, and by June 1997 there were 323 orders from 25 airlines.

Initially the JAA were more reticent about granting Early-ETOPS to European operators and restricted operations to 120-minutes ETOPS until in service reliability was proven.

The first year of operation resulted in zero IFSDs with all three engine operators; however, the honeymoon period was short lived. The Pratt & Whitney PW4077 reached a peak 12-month rolling average engine IFSDR of 0.018/1,000 hours in October 1996. The General Electric GE90-77B reached a peak of 0.021 for one month in July 1998 and the Rolls-Royce Trent reached a peak of 0.016 in December 1997. Although the in-flight shutdown rates stayed within the allowable 0.02/1,000 hour required for 180-minute ETOPS, significant design problems were discovered on each engine type after ETOPS approval.

This raised concerns with the FAA regarding the efficacy of the ETOPS process, since the failure modes had not been apparent in the enhanced test programme. Some failure modes had the potential to result in in-flight shutdowns had they occurred under different circumstances, or had they not been detected during maintenance for unassociated reasons. The FAA was concerned that had every one of those events resulted in an engine in-flight shutdown, the resulting IFSDRs for each engine type could have been significantly higher, and exceeded the 0.02/1,000 hours required for 180- minutes approval.

Boeing, the engine manufacturers, the FAA, and other regulatory authorities worked together to prevent additional in-flight occurrences of these failure modes. The actual in-flight shutdown rates prove that these early in-service problems were successfully managed to maintain the safety of Boeing 777 ETOPS operations worldwide. The FAA concerns were underlined due to the initial in-service experience with the Boeing 737 NG (New Generation); NG included the 737-600, 700, 800 and 900, all of which were powered by variants if the CFM 56 engine. The 737 NG had been tested to a set of conditions similar to those in 25 ANM -84, which Boeing was proposing as the basis for early 180-minute ETOPS approval for the extended range 777-200 ER.

The Model 737-700 was the first variant of the 737NG to enter service in December 1997 and the fleet had reached 15,000 hours of operation during April 1998. At that time, there had been no IFSDs in service, however, on 9 May 1998, before the FAA had completed its assessment of the airplane for ETOPS approval, the first IFSD occurred, followed by a second later that month such that the fleet exceeded the accepted 120-minute ETOPS standard of 0.05 IFSDs/1,000 hours. Three further IFSDs occurred in the following month, followed by another in July, making a total of six in a three-month period. This raised the IFSDR to 0.085/1,000 hours which clearly did not meet the minimum standard for ETOPS type design approval.

The six engine IFSDs were caused by three different failure root causes. Boeing and the engine manufacturer CMFI undertook aggressive actions to correct each of these design problems as they occurred. Due to the rapid accumulation of flying hours generated by the Boeing 737 NG fleet of aircraft during this period new ETOPS reportable events occurred faster than the known problems could be corrected. This delayed FAA consideration of the 737-700 for ETOPS approval until the problems were brought under control. Without any further IFSDs occurring the high rate of accumulation of flight hours had a beneficial effect, in that the IFSDR decreased rapidly and was within the ETOPS type design approval standard by the end of 1998. The FAA approved the 737-600/-700/-800 (737NG) for 120-minute ETOPS approximately one year after EIS with over 300,000 engine-hours of service experience and an in-flight shutdown rate of 0.020/1000 hours.

Analyses of these events from the 737 and 777 service operation, in comparison with the enhanced testing and procedures, led the FAA and Boeing to the conclusion that some modifications were required to the special conditions laid out in 25 ANM-74. The FAA concluded that the following five elements, when combined, provide an acceptable substitute for actual airline service experience.

1. Design for reliability.
2. Lessons learned.
3. Test requirements.
4. Demonstrated reliability.
5. Problem tracking system.

The main conclusions were that post-test inspections of the 3,000 cycle tests had not been thorough enough, with the consequence that potential

IFSD events were overlooked. The importance of the post-test strip-down analyses was emphasised as follows:

"At completion of the engine demonstration test, the engine and airplane nacelle test hardware must undergo a complete teardown inspection. This inspection must be conducted in a manner to identify abnormal conditions that could become potential sources of engine IFSD. An analysis of any abnormal conditions found must consider the possible consequences of similar occurrences in service to determine if they may become sources of engine IFSDs, power loss, or inability to control engine thrust. Any potential sources of engine IFSD identified must be corrected to ensure the effectiveness of the Early ETOPS process."

It was agreed that the 1,000 cycle flight test incorporating an ex-development cyclic test engine, was less likely to highlight problems that had not already been discovered by bench testing. In hindsight it was concluded that more benefit would come from testing the engines for longer periods at cruise conditions, particularly under the one engine failed scenario over the maximum diversion time conditions. Accordingly, the special conditions were amended to include this more representative approach.

A post EIS review of IFSDs and incidents conducted by the FAA, Boeing and the respective engine manufacturers on both the 737 and 777, and the recovery actions taken to recover the IFSDR, highlighted the importance of the Reliability Assessment and Problem Tracking System and this was emphasised in the Special Conditions amendment.

The early Early-ETOPS requirements are now defined in FAR regulation 25.1535 appendix K25.2.2(d) and outlined in Advisory Circular AC33 201-1.

The Boeing 777 outsold the A340 after its launch. In 2002, with flagging sales and prior to the introduction of its larger A340-600, in an apparent attempt to undermine the success of the 777, Airbus unwisely placed an advertisement at the Farnborough Air Show claiming that 4-engine aircraft were safer than 2-engine aircraft over long range routes. A colossal billboard at the edge of a runway read, "*A340 - 4 engines 4 long haul*." Similar full page advertisements appeared in air show magazines and daily London newspapers. The advert was essentially undermining the ETOPS process, which was surprising as their own A300, A310, A320 and A330 were operating under ETOPS rules. This was not the first time Airbus had created controversy with its advertising strategy. Apparently in 1999 they released and then pulled from circulation an advertisement showing an A340 flying over a vast, stormy ocean, with a caption stating, *"If you're over the middle of the Pacific, you want to be in the middle of four engines."* Boeing, airline representatives and engine manufacturers were quick to come to the defence of the 777, and as a result of the negative reaction Airbus withdrew the advert. Surprisingly Airbus A340 sales picked up slightly after the introduction of the -600 version but the improvement was short lived.

Not surprisingly Boeing soon issued data to counter the Airbus argument that the A340 was safer than the 777. By 2002 there had been approximately 3,000,000 ETOPS flights logged since 1985, and over 1,000 ETOPS flights were taking off every day, during which time no twin-jet had been lost during the ETOPS portion of the flight, i.e. the portion when the airplane is farthest from an alternate airport. At the time, out of over 300,000 flights, the Boeing 777 had experienced only one diversion for an in-flight engine shutdown during an ETOPS flight. In countering the Airbus claim, Boeing took the opportunity to cite the advantages of the 777 over the A340; namely 20 percent lower fuel burn which equated to 18 percent lower fuel cost per passenger. Boeing data also showed that the 777 had half as many air turn-backs and diversions/1,000 departures as the A340.

It soon became apparent from service experience that the Early-ETOPS process had a beneficial effect on engine reliability, with IFSDRs being further reduced to 0.01/1,000 hours and in many cases significantly below that. The longest ETOPS diversion recorded occurred on a United Airlines 777-200 powered by Pratt & Whitney 4077 engines during a 180-minutes ETOPS flight over the Pacific from Auckland New Zealand to Los Angeles. The aircraft, with 255 passengers on board, was well past the midway point to Hawaii when the number two engine was shut down due to high oil temperature; the aircraft was diverted to Kona and flew for 192 minutes before landing safely. Flight time exceeded the planned 180 minutes due to strong head winds over the Pacific. This was only the third recorded diversion of a 777 in 400,000 ETOPS flights, before this event the United Airlines 777 fleet had recorded a total of 16 in-flight shutdowns during all phases of flight since May 1995, during which time it had flown a total of 2.3 million hours, recording an IFSDR of 0.0021/1,000 hours.

There seems little doubt that the lower IFSDRs enjoyed by the Boeing 777 were a direct result of the enhanced design and development procedures adopted during the Early-ETOPS process, further developed in the rigorous follow up after EIS. In 2007 the RR Trent 877 was recorded as having an IFSDR of 0.0048/1,000 hours and in 2014 GE advertised that the GE90 was enjoying an IFSDR of 0.001/1,000 hours after accumulating 40 million flight hours since its introduction in 1995. It seems inevitable that the experience of this new process would be fed into future aircraft and engine design and development programmes, irrespective of whether they were to be approved for ETOPS via the Early-ETOPS route, or by the service experience route, such that the benefits would accrue to all aircraft and engine types.

In 1994 Boeing introduced the –300 ER (Extended range) version of the 777, having a maximum range of 9,000 miles, increased seating capacity to 550 maximum single-class, and a maximum takeoff weight of 775,000lb. The consequent increase in weight required an increase in thrust to 115,000lbf and Boeing selected the GE 115-B engine as sole power plant for the aircraft.

The GE 90-115B is the most powerful aero-engine in service to date. With a bypass ratio (BPR) of 9:1 the engine delivers 115,000lbf maximum takeoff thrust; the 128-inch diameter fan and four Low Pressure Booster (LPB) stages are driven by six stages of Low Pressure Turbine (LPT).

Fig 89: GE 90-115B

The High Pressure Compressor (HPC) has 9 stages driven by two High Pressure Turbine (HPT) stages; the compression system delivers an Overall Pressure Ratio (OPR) of 42:1, the highest in existence at the time. To put it into context, the nacelle diameter is similar to that of the Boeing 737 fuselage. It is interesting to compare the engine characteristics with the GE 80C-2 that powered the Boeing 767 twenty years before (see Table 4, Appendix 2). The BPR increased by 78 percent, thrust increase was 95 percent for a mass flow increase of about 70 percent, OPR increased by 38 percent even though the total number of compression stages reduced by five, reflecting the improvement in technology and analytical tools over the period. In spite of all this, the engine demonstrated significantly improved IFSDR and dispatch reliability (at least 99.98 percent of departures being achieved within 15 minutes of schedule), which is a testament to the improved design, test and maintenance procedures that evolved through the ETOPS process. The GE 90-115 B currently has the lowest IFSDR on the 777. However, as it is exclusive to the extended range aircraft, the IFSDR per flight is probably similar to the other comparable engines.

With IFSDRs of less than 0.01/1,000 hours, it is not surprising that the aircraft manufacturers and airline operators wished to increase the permitted diversion times so as to optimise their operations on those routes where 180 minutes was insufficient to permit the most direct route to the destination; most of these routes were in the South Pacific, Indian Ocean and Polar Regions. For these routes, an allowance of a 330-minute diversion time would satisfy most requirements, whilst still meeting the extremely improbable double engine failure scenario. At the same time, in the interests of safety the FAA was considering introducing some of the elements of the ETOPS process to 3- and 4-engine aircraft.

In June 2000, following 15 years of successful ETOPS operation, the FAA tasked the Aviation Rulemaking Advisory Committee (ARAC) to set up a working group (WG) to review the experience of ETOPS and to provide advice and recommendations for the future. The WG was set the following objectives:

1. Review the existing policies and requirements found in the existing ETOPS Advisory Circular (AC) 120-42A, and other applicable ETOPS special conditions, policy memorandums and notices for certification and operational regulations and guidance material for ETOPS approvals up to 180 minutes.
2. Develop comprehensive ETOPS airworthiness standards for FAA

Regulation 14 CFR parts 25, 33, 121, and 135, as appropriate, to codify the existing policies and practices.
3. Develop ETOPS requirements for operations in excess of 180 minutes up to whatever extent that may be justified, and develop those requirements such that incremental approvals up to a maximum may be approved.
4. Develop standardised requirements for extended range operations for all airplanes, regardless of the number of engines, including all turbojet and turbo-propeller commercial twin-engine airplanes (business jets), excluding reciprocating engine powered commercial airplanes and establish criteria for diversion times up to 180 minutes that is consistent with existing ETOPS policy and procedures. It should also develop criteria for diversion times beyond 180 minutes that is consistent with the ETOPS criteria developed by the working group.
5. Develop additional guidance and/or advisory material as the ARAC finds appropriate.
6. Harmonise such standardised requirements across national boundaries and regulatory bodies.

The ARAC issued its conclusions to the FAA in October 2002, which included a number of recommendations for change aimed at improving the process. Additionally, recommendations were made for ETOPS operations beyond 180 minutes for twin-engine aircraft. During the review it was recognized that the safety of 3- and 4-engine aircraft could be improved by applying some of the practices developed for twin-engine aircraft e.g. pre-flight planning for diversion, ensuring adequate fuel reserves in the event of a diversion being required. As a consequence, the new rules were extended to include requirements for 3- and 4-engine aircraft operating in excess of 180 minutes from a diversion airport. The acronym ETOPS was changed from Extended Twin-Engine Operations (ETOPS) to Extended Operations (EROPS). The revised regulations were included in FAR25 from January 2007; Advisory Circular AC 120-42A was replaced by 120-42B to reflect the changes. A summary of the main changes is as follows:

- New requirements were issued for expanded ETOPS beyond 180 minutes. These included passenger recovery plans, particularly for diversions in polar regions, satellite communications, fire suppression capability within the cargo holds to a timescale 15 minutes longer than the maximum envisaged diversion time, enhanced aircraft

certification requirements including fuel system pressure and flow requirements, low fuel quantity alerting system, engine oil tank design to prevent oil loss due to mal-assembled filler caps and IFSDRs of less than 0.01/1,000 hours.
- For 3- and 4-engine aircraft the ETOPS operation programme, training for flight crews and dispatch, cargo fire suppression, critical fuel scenarios and passenger recovery plans apply. A timescale of eight years was given for compliance with the new type design requirements, and six years for compliance with the new rules on cargo hold fire suppression.

In addition, the above the qualification requirements for operation of ETOPS beyond 180 minutes are as follows:
- For ETOPS beyond 180 minutes, up to and Including 240 minutes, the FAA grants approval only to certificate holders with existing 180-minute ETOPS operating authority for the airplane-engine combination to be operated in the application. There is no minimum in-service time requirement for the 180-minute ETOPS operator requesting ETOPS approval beyond 180 minutes.
- For ETOPS beyond 240 minutes the approval is only granted to operators of two-engine airplanes between specific pairs of cities. The certificate holder must have been operating at 180 minutes or greater ETOPS authority for at least 24 consecutive months, of which at least 12 consecutive months must be at 240 minute ETOPS authority with the airplane-engine combination in the application.
- There are no minimum in-service experience criteria for certificate holders requesting ETOPS beyond 180 minutes for operations with passenger carrying aircraft having more than two engines. Those applicants will request approval under the accelerated ETOPS method.

In November 2009 the Airbus A330-300 became the first aircraft to receive 240-minutes ETOPS type approval, which it offered to its customers as an optional requirement, but it was Air New Zealand who completed the world's first commercial 240-minute ETOPS flight in December 2011 between Los Angeles and New Zealand with a Boeing 777-300 ER.

In the same month, Boeing announced that they had received type-design approval from the FAA for up to 330-minute extended operations for its GE 90 powered 777 fleet, with approval for the RR and Pratt & Whitney engine variants due shortly after. Air New Zealand was also the first airline to take up the 330-minute option.

The latest generation of twin-engine long range aircraft have been designed with extended range ETOPS up to and including 370 minutes as an option for those airlines whose route structures would benefit from such long diversion times. The Boeing 787 Dreamliner was designed as a more efficient replacement for the 767. The aircraft introduced several new technologies, including extensive use of carbon fibre material in place of aluminium alloy, enabling significant weight reduction to be achieved.

Additionally, the use of carbon fibre in the fuselage has enabled the cabin pressure to be increased to the equivalent of 6,000 feet altitude, making it a more comfortable environment for passengers. Cabin air pressurisation is provided by electrically driven compressors, rather than traditional engine-bleed air, thereby eliminating the need to cool heated air before it enters the cabin. The cabin's humidity is programmable, based on the number of passengers carried, and allows 15 percent humidity settings instead of the 4 percent found in previous aircraft. The cabin air-conditioning system improves air quality by removing ozone from outside air, and in addition to standard filtration to remove airborne particles it uses a gaseous filtration system to remove odours, irritants and gaseous contaminants, as well as viruses, bacteria and allergens. The bleed-less engine cabin air system also allows the 787 air to avoid the risk

Fig 90: A British Airways Boeing 787-8 Dreamliner

of oil fumes and toxins which may be present to some degree in conventional air bleed systems.

The first variant, the 787-8, was aimed at replacing the 767-200ER and 300ER. All Nippon Airways (ANA) was the first customer and on 26 October 2011 it flew its first commercial flight from Tokyo Narita to Hong Kong. Tickets for the flight were sold in an online auction. The highest bidder paid $34,000 for a seat; quite a bit more than the $400 paid 97 years earlier by the first fare-paying passenger on the Benoist XIV. An ANA 787 flew its first commercial long-haul flight on 12 January 2012 from Haneda to Frankfurt.

The second variant is the 787-9, a stretched version of the 787-8, introduced to compete with the Airbus A330-200 and 300, and as a replacement for the ageing 747- 400ER. The 787-9 began commercial service with ANA on 7 August 2014. Currently the longest non-stop flight is operated by United Airlines between Los Angeles and Melbourne. A further stretched version, the 787-10, is currently in the process of development.

The engines selected to power the 787 are the RR Trent 1000 and the GE GEnx-1B, each rated at 64,000 to 71,000 lbf thrust. The Trent 1000 was the launch engine and is the fifth in the series of Trent engines currently in service on the A330, A350 XWB, A380, Boeing 777 and 787. The fan diameter is 112 inches and BPR is 10.8:1; the core engine configuration is similar to that of the Trent 900 that powers the Airbus A380. The GEnx is a smaller version of the GE 90 with a BPR of 9.6:1.

The 787-8 was initially approved for 180-minutes ETOPS at EIS; although there have been isolated IFSDs on both engine types the rate has remained within the 0.01/1,000 hours required for extended range ETOPS. In May 2014 the 787-8 was granted 330-minutes ETOPS approval as an option for certain customers. This extended diversion time opens up the possibility of more economical operations by the twin-jet between Sydney and Santiago, and Sydney and Johannesburg, with the possible replacement of 747-400s by 787s on these routes. The 787 flight from Johannesburg to Sydney is about four hours shorter as a 330-minute ETOPS operation than it would be as a 180-minute ETOPS operation.

In December 2004 Airbus launched the design of a new aircraft designated the A350 XWB (Extra Wide Body). The A350 was initially planned to be a 250- to 300-seat twin-engine wide-body aircraft derived from the design of the existing A330. Under this plan, the A350 would have modified wings and new engines, while sharing the same fuselage cross-section as its predecessor. Two versions were planned to compete directly with the Boeing 787-9, and 777-200ER. The proposal was not favourably received by potential customers, and consequently Airbus decided to embark on a more adventurous and competitive design. The use of the A330 fuselage was abandoned and a wider fuselage was designed, being slightly wider than the Boeing 787. Like the 787, carbon composite material was used in the fuselage and wing design and the cabin pressure was increased. The aircraft was renamed A350 XWB and was slightly larger than the 787.

Three versions were proposed. The A350-800 would compete with the 787-9 and directly replace the Airbus A330-200, the A350-900 would compete with the Boeing 777-200ER and replace the Airbus A340-300, and the A350-1000 would compete with the Boeing 777-300ER and replace the A340-600.

Fig 91: Airbus A350-900 XWB

Airbus selected Rolls-Royce as the sole engine supplier and a new version of the Trent, designated Trent XWB, was designed with a thrust range of 79,000 to 97,000 lbf for the three variants. The engine is advertised as the most fuel efficient engine produced to date. In addition to the introduction of a number of technical improvements, the main configuration change is the addition of a second Intermediate Pressure Turbine (IPT) stage, enabling higher pressure ratio to be achieved in the Intermediate Pressure Compressor (IPC) without degrading IPT efficiency.

The fan diameter is 118 inches making it the largest of any of the Trent

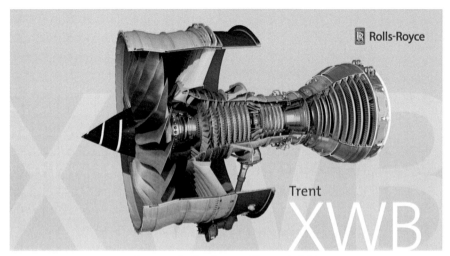

Fig 92: Rolls-Royce Trent XWB

engines; BPR is 9:1. The engine has an 8-stage IPC, 6-stage HPC, single-stage HPT, 2-stage IPT and 6-stage LPT, the OPR is 52:1

The A350-9 entered service with Qatar Airways on 14 January 2015, flying from Doha to Frankfurt.

In October 2014 the European Aviation Safety Agency (EASA) approved the new A350-900 airliner for ETOPS beyond 180-minutes diversion time, making the A350 XWB the first new aircraft type ever to receive such a level of ETOPS approval prior to EIS. The approval, which includes 180-minutes ETOPS in the basic specification, also includes provisions for 300-minute ETOPS and 370-minute ETOPS depending on individual operator selection. The latter option extends the diversion distance up to an unprecedented 2,875 miles, enabling A350 operators to serve new direct non-limiting routings compared with a 180-minute ETOPS diversion time.

The ETOPS 370-minute option will be of particular benefit for new direct southern routes, such as between Australia, South Africa and South America. The ETOPS 300-minute option will facilitate more efficient transoceanic routes across the North and Mid-Pacific, such as from South East Asia and Australasia to the USA. Operators flying on existing routes under the 180-minute rule will be able to fly a straighter and consequently quicker and more fuel efficient path, and also have access to more and possibly better equipped en route diversion airports if needed. Airbus had originally aimed for 420-minutes approval and it would not be surprising if this were eventually agreed.

With these extended diversion times and larger twin-engine aircraft, the opportunities for 4-engine aircraft are limited; the A380 is effectively the only 4-engine powered civil aircraft in production. Replacement of an aircraft the size of the A380 with a twin-engine variant seems highly unlikely at present as engines of approximately 150,000 lbf thrust would be required. Such an engine would probably require a fan diameter approaching 150 inches which would require a very large diameter multi-stage LPT to drive it. The more likely scenario is that the Boeing and Airbus will stretch the 787 and A350 taking advantage of the carbon fibre construction to minimize weight, and the engine manufacturers will develop the geared-fan concept to offset increases in weight due to higher fan diameters.

Pratt & Whitney have already produced the PW 1100G geared-fan engine for the Airbus A320 NEO (New Engine Option) to compete with the more conventional CFM LEAP 1-A.

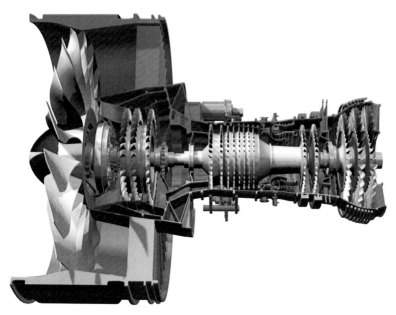

Fig 93: P&W 1100G Geared Fan

The geared-fan has a reduction gear ratio of about 3:1 allowing the LPC and LPT to run at their optimum speed. The LPC and LPT have fewer stages and are reduced in diameter, enabled by the increase in LP Rotor speed. The geared-fan arrangement is not a derivative of a prior engine, being a hybrid of a turboprop and turbofan. It will be interesting to see what approach the authorities take regarding ETOPS at EIS for the PW1100 powered A320 NEO.

Rolls-Royce has announced its intention to develop a geared-turbofan in the higher thrust range, with the core engine being based on its unique 3-shaft arrangement. It seems likely that all three engine manufacturers may ultimately finish up with variations of the same concept.

Conclusion

With the approval of extended diversion times, and the introduction of larger, long range twin-engine aircraft, ETOPS is fast becoming the standard form of travel. This is reassuring for passengers since they can travel in the knowledge that the flight route has been preplanned to cater for the unlikely event of an engine shutdown. Engine fuel reserves exist for such an eventuality taking account of the revised flight altitude and cruising speed with one engine operative; cargo fire suppression has been provided to cater for at least 15 minutes more than the maximum possible diversion time; conditions at the potential diversion airport have been considered and are checked periodically throughout the flight; satellite communication is a mandatory requirement for any ETOPS flight; maintenance procedures have been carried out to prevent common cause human errors; the fleet engine IFSDRs are within those specified within the ETOPS approval consistent with ensuring the probability of double engine failure is less than one per billion flight hours; reliability procedures are in place to prevent the IFSDR exceeding the limits specified in the regulations.

For extended range operations beyond 180 minutes the rate is 0.01/1,000 hours (1×10^{-5}), the lower the rate the more important it is to react to any failures that may occur, there is clearly an incentive on the airlines, aircraft and engine manufacturers to react promptly to any IFSD events to prevent a reduction in diversion time being imposed.

On-line engine condition monitoring can track the engine health and schedule removals prior to age-related problems occurring, information can also be used to diagnose problems as they occur. Three and four-engine aircraft travelling more than 180 minutes from a suitable diversion airport are also adopting some of the ETOPS procedures, particularly pre-flight planning, fuel reserve provisioning for potential diversions, and cargo fire suppression.

There is no doubt that the ETOPS process has had a significant beneficial effect on aircraft safety and economics and is here to stay. However, there is no room for complacency, as the number of ETOPS missions will continue to increase as the modern twins take over from the ageing Boeing 747s, and advantage is taken of the latest extensions to diversion times.

Appendix 1

Notes

- Turboprop Engines
- Russian Aircraft
- Reliability
- Jet Engine Trends 1940-2015
- The Demise of the British Civil Aircraft Industry
- Conversion Factors
- Abbreviations

Turboprop Engines

A turboprop is a gas turbine engine that drives an aircraft propeller. In contrast to a turbojet, the engine's exhaust gases do not contain enough energy to create significant thrust, since almost all of the engine's power is used to drive the propeller. One of the best known turboprop engines in civil application was the Rolls-Royce Dart. First introduced in 1948, the Rolls-Royce Dart turboprop combined the power of jet propulsion with the efficiency of propellers, it was widely used in the first generation of turboprop-powered aircraft, including the British Vickers Viscount and the Dutch Fokker F-27. The Dart enabled these and the other new airliners to reduce airline operating costs and bring greater speed and comfort to passengers traveling on short-to-medium length routes.

Turboprop engines are generally used on small subsonic aircraft, being most efficient at flight speeds below 450 mph because the air velocity of both the propeller and exhaust is relatively low. Due to the high price of turboprop engines, they are mostly used where high-performance short-takeoff and landing capability and efficiency at modest flight speeds are required. The most common application of turboprop engines in civilian aviation is in small commuter aircraft, where their greater power and reliability than piston engines offsets their higher initial cost and fuel consumption. Turboprop airliners now operate at near the same speed as small turbofan-powered aircraft, but burn two-thirds of the fuel per passenger. However, compared to a turbojet, which can fly at high altitude for enhanced speed and fuel efficiency, a propeller aircraft has a much lower ceiling.

Fig 94: Rolls-Royce Dart
2-stage centrifugal flow compressor, 7 combustion chambers, 3-stage axial-flow turbine. Power: 1,815 HP at 15,000 rpm, weight: 1,250 lb, SFC: 0.58 lb/hp/hr., OPR 5.6:1

The Bristol Aircraft Company introduced a long range turboprop, the Bristol Britannia Series 310 which began transatlantic flights with BOAC from London to New York in 1957. The aircraft cruised at 350 mph and had a range of 4,400 miles, and seating for 139 coach-class passengers; however, it was uncompetitive on the transatlantic routes with the bigger

Fig 95: Bristol Britannia

Fig 96: Bristol Proteus Turboprop

and faster Boeing 707 and Douglas DC8 aircraft that were introduced a year later. The Britannia was powered by four Bristol Proteus Turboprops.

A Proteus 755 engine delivering 4,250 HP, powered Sir Donald Campbell to a land speed record of 403 mph in his Bluebird car.

Turboprop powered aircraft were not included in the main text as they had minimal effect on the development of ETOPS.

Russian Aircraft

Russia has its own self-contained aircraft and engine industry. They have produced some noticeable twin-engine civil aircraft including the Tupolev TU 204 and 214, and more recently the Sukhoi Superjet 100. Details are fairly sparse and sales have been mainly to eastern bloc airlines and in no way compare with the volume of sales of Boeing and Airbus. They have had no influence on the evolution of ETOPS.

Reliability

The reliability figures for IFSDRs derived from the formula in Chapter 4 are the probability of a duel-engine failure/flight hour due to a single engine cause; to arrive at the probability of failure/flight it is necessary to multiply by the flight duration. Assuming a flight duration of 20 hours, and a diversion time of 370 minutes (6.17 hours), and the maximum permissible in-service reliability rate of 0.01/1,000 hours (1×10^{-5}) for extended range ETOPS, the probability of failure/flight hour is 0.9×10^{-9}, i.e. just within the extremely improbable requirement. The probability of failure/flight for a journey time of 20 hours is therefore 1.8×10^{-8}. This compares with an all causes World hull loss accident rate for scheduled commercial operation between 2004 and 2013 of 0.61×10^{-6}, i.e. 340 times better. In fact, the probability estimate is somewhat pessimistic since it assumes that the first engine failure occurs at the worst point in flight, also the typical IFSDR being achieved by ETOPS approved twin-engine aircraft is less than half the IFSDR required for extended range ETOPS approval. Also the evidence from the latest generation of high bypass ratio twins is that cruise IFSDR is nearer to one third of the overall rate rather than one half, which if maintained will result in a further reduction in dual failure probability.

Duel-engine failures due to a single cause, e.g. multiple bird strike, volcanic ash ingestion, complete loss of fuel etc., are not included in the statistics as they are unrelated to basic engine reliability.

Jet Engine Trends 1940-2015

The graph opposite illustrates how the jet engine has developed from 1940 to 2015.

Thrust has increased by a factor of ten from 10,000 lbf for the turbojet to 115,000 lbf for the high bypass turbofan (GE90-115B), the change in slope of the graph coincides with the introduction of the high BPR turbofans for the Boeing 777 and Airbus A330. The thrust-to-weight ratio has increased from about 2:1 to 6.4:1 but the improvement has leveled off with the introduction of the high BPR turbofans. The initial steady improvement can be attributed to improvements in aerodynamics with higher OPRs achieved with fewer stages (see OPR/Stage), the use of higher strength lower density material, and better analytical techniques due to the introduction of computers.

The high thrust of the high BPR turbofans has been achieved by increasing fan diameters, the highest being the 128 inch GE90-115B. The aerodynamic limitation on fan tip speed (approx. 1450ft/sec) necessitates an increase in diameter of the LPC booster stages and LP Turbine to maintain acceptable blade mean speed for optimum component efficiency. An increase in the number of LPT stages exacerbates the problem resulting in an adverse effect on engine weight.

OPR has increased by a factor of ten, whilst the number of compression

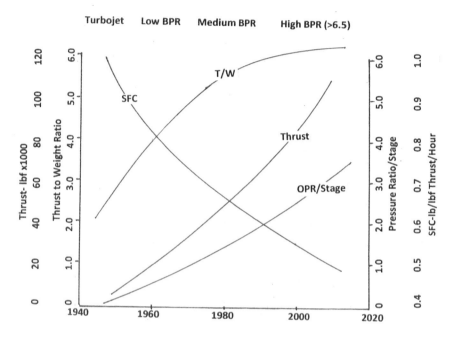

Fig 97: Jet Engine Performance Trends with Time

stages in the axial-flow configuration has stayed broadly the same. This is attributable to the availability of 3D aerodynamic analyses programmes and powerful computer processing capability, high strength high temperature alloys facilitating increased compressor delivery temperatures, and the introduction of active tip clearance control facilitated by sophisticated thermal analyses programmes. OPR/stage has increased from 0.5 to 3.5.

Specific fuel consumption (SFC) has halved from 1.10 lb/lbf thrust/hour to 0.52. Much of the early improvement is attributable to the introduction of the bypass turbofan and the trend has continued as BPR and OPR have increased. The trend is still downwards and looks set to fall below 0.5 when the next generation of geared fans with even higher BPRs and OPRs are introduced.

The Demise of the British Civil Aircraft Industry

I could not help wondering during the course of researching this narrative how the demise of the British Aircraft Industry came about. Post WW2 Britain had a strong but fragmented aircraft industry which has declined progressively over the years. In 2006 BAe Systems sold its

Fig 98: BAe 146-The Last British Airliner

20 percent share in Airbus to the French, Spanish and German group EADS (European Aeronautic Defence and Space Company), and in so doing became a subcontractor to Airbus.

The last British designed and produced civil airliner was the BAe 146 regional jet, later re-named the Avro RJ, which entered service in 1983 with Dan Air. A total of 387 were built and production ceased in 2002 ending an 80-year era of civil aircraft production.

There are a number of reasons which have contributed to this decline, not least of which is poor decision making by successive governments as follows: -

If the Air Ministry had had the vision to give Sir Frank Whittle the backing to develop his aero-engine at the beginning of 1930, the Meteor jet fighter might have been available at the beginning of WW2 instead of the end, which could have led to a totally different outcome.

Failure to make Whittle's patent secret, and to renew it in 1935, gave Germany valuable information which probably accelerated progress on their own jet engine design. Following the first flight of the Whittle W1 powered Gloster E28/39, the Americans became interested and asked for the details and an engine. At the end of 1941 a Power Jets team and a W1X engine were flown to Washington to enable General Electric to examine it and begin construction.

This kick-started US Jet engine development, ultimately to the cost of

Fig 99: Bell B59-B Airacomet

the British industry, the US also obtained the details of the German jet engine developments at the end of WW2, which further added to their database.

The Americans developed the idea and their Bell XP-59A Airacomet was airborne on 2 October 1942.

In 1942 the US government made an agreement with the UK that they would concentrate on the production of transport and civil aircraft whilst the UK concentrated on fighters and bombers. This resulted in the US developing the large radial engines required for the large transport aircraft, which were quickly converted to civil airliners after the war. The UK had no aircraft of comparable size to compete on the intercontinental routes.

The catastrophic failures of the De Havilland Comet 1 were a major setback to the British industry, enabling the US to re-establish a monopoly on transcontinental flights with the Boeing-707 and Douglas DC-8.

The fragmented nature of the post war British aircraft manufacturers meant that there was no single company large enough to compete with the US companies of Boeing, Douglas and Lockheed. It was 1960 before this was whittled down to two main companies, the British Aircraft Corporation (BAC) and Hawker Siddeley, both of whom had moderate success, but not in the large US market. It was 1977 before these companies were amalgamated in to one entity as British Aerospace (BAe).

The two main British airlines after the war were the British Overseas Airways Corporation (BOAC) and British European Airways (BEA), both these airlines were nationalised. In 1974 these airlines were amalgamated, along with some minor airlines, to form British Airways which was also nationalised and remained so until it was privatised in 1984. BAe was nationalised at its formation in 1977 and remained so until the government sold 51.7 percent of its shares in 1981 and the remainder in 1985. Rolls-Royce went into liquidation in 1971 and became nationalised to protect its military products; it was eventually re-privatised in 1987. The effect of this nationalisation was that too many decisions were based on political issues as opposed to commercial considerations to the detriment of the industry.

In 1969 the British government withdrew from the proposed Airbus consortium in favour of the Concorde project. The claim was that they could not afford to fund both, the UK's then Labour minister of technology, Anthony Wedgwood Benn, said the government was "not satisfied that it was a good investment to go ahead with the project at this stage". The UK was being asked to find one-third of the estimated £180 million development costs for the 'A-300', which at the time was attracting little interest from potential customers; as events turned out they backed the wrong horse. In the meantime, the Concorde's French partner, Sud Aviation, pursued a parallel course by becoming a founder member of Airbus Industrie with Germany; the result being that Toulouse Blagnac Airport, the home of Sud Aviation, is now the most important aerospace city in Europe.

Additionally, insufficient funding was available to enable Rolls-Royce to pursue both the L1011 and proposed Airbus A300, such that RR had to withdraw from the latter. It was another twenty years before an RR engine was selected to power an Airbus aircraft.

As a result of these decisions Britain lost the opportunity to become a major shareholder in Airbus Industrie to the detriment of its civil aircraft industry.

The licensing of engines to other manufacturers was commonplace until the mid-1950s. The financial benefits may have seemed worthwhile at the time but probably were detrimental in the long term. The American companies benefited from licence agreements with both British and French companies, which was beneficial to the development of both their piston and jet engine industry.

Units, Conversion Factors and Abbreviations

Units and Conversion Factors

Imperial units have been used throughout the text. The figures used to describe the relative engine and aircraft features and performance have been extracted from various sources and cannot be guaranteed as definitive, however they are accurate enough for comparative purposes and to illustrate trends in engine and aircraft development.

Standard conversion factors to Metric units are given below:
1 mile = 0.869 nautical miles.
1 inch = 2.54 centimetres
1lb = 0.453 kg
1,000 lbf = 4.448 kN
1lb/in² = 0.0069 MPa

Abbreviations

ANA - All Nippon Airways.
APU - Auxiliary Power Unit
ARAC - Aviation Rulemaking Advisory Committee.
BAC - British Aircraft Corporation.
Bae - British Aerospace.
BEA - British European Airways.
BOAC - British Overseas Airways Corporation.
BPR - Bypass Ratio (Ratio of mass flow of compressed air bypassing the combustion process to that entering the combustion process).
DH - De Havilland.
ER - Extended Range
EADS - European Aeronautic Defence and Space Company.
EASA - European Aviation Safety Agency.
ETOPS - Extended Twin-Engine Operations.
EIS - Entry into Service.
FAA - Federal Aviation Administration.
GE - General Electric.
HPC - High Pressure Compressor.

HPT - High Pressure Turbine.
HP - Horsepower.
ICAO - International Civil Airworthiness Organisation.
IFSD - In-flight shutdown.
IFSDR - In-flight shutdown rate.
IFSDRs - In-flight shutdown rates.
IPC - Intermediate Pressure Compressor.
IPT - Intermediate Pressure Turbine.
JAA - Joint Airworthiness Authority.
LPC - Low Pressure Compressor.
LPT - Low Pressure Turbine.
lbf - Pounds Force.
OPR - Overall Pressure Ratio (The ratio of the compression system delivery pressure over the inlet pressure).
Pan Am - Pan American Airways.
P&W - Pratt and Whitney.
PRT - Power Recovery Turbine.
psi - Pounds per square inch.
RR - Rolls-Royce.
rpm - revolutions per minute.
SFC - Specific Fuel Consumption.
TWA - Trans World Airways.
TET - Turbine Entry Temperature.
TDC - Top Dead Centre.
UK - United Kingdom.
USA - United States of America.
US - United States .
WW1 - World War 1.
WW2 - World War 2.

Appendix 2

Tables

Table 1
The table summarises the approximate performance details of the several engines which had similar geometric configurations, being single spool, centrifugal-flow compressors with straight through combustors and single stage turbines, all of which were derived from the original Whittle engine.

Table 2
The table shows the beneficial effects of increased mass flow on thrust-to-weight ratio and increased pressure ratio on SFC.

Table 3
The table shows the comparative performances of a selection of the competing low bypass ratio engines available in the early 1960s.

Table 4
The table shows the comparative features and performance of the initial high bypass ratio turbofans and their developed versions that made twin engine operation achievable.

Appendix 2: Tables

Table 1

The table summarises the approximate performance details of the several engines which had similar geometric configurations, being single spool, centrifugal-flow compressors with straight through combustors and single stage turbines, all of which were derived from the original Whittle engine.

Data	RR Derwent Mk5	DH Goblin	Alisson J33	RR Nene	DH Ghost	Klimov VK-1	P&W J48 P8 A	Hispano Suiza Verdon
Diameter	43	50	50.5	49.5	53	51	50.5	50
Length	88.5	107	107	96.8	121	102	109.8	103.2
Weight	1250	1550	1820	1600	2218	1922	2090	2061
Thrust	4000	3000	4600	5000	5125	5955	7250	7700
RPM	15000	10200	11500	12300	10250	NK	NK	11000
T/W	3.2	1.9	2.53	3.2	2.25	3.1	3.48	3.73
OPR	3.9/1	3.3	4.1	4.5	4.6	NK	NK	4.9
Mass Flow	NK	60	87	90	87	NK	115	132
SFC	1.28	1.18	1.14	1.6	1.02	1.07	1.14	1.1

In Tables 1&2: Diameter & Length in inches, Weight in lb, Thrust in lbf, T/W is thrust-to-weight ratio, OPR is overall pressure ratio, Mass Flow lb/sec, SFC in lb/hour/lbf thrust.

Table 2

The table shows the beneficial effects of increased mass flow on thrust-to-weight ratio and increased pressure ratio on SFC.

Single-Spool, Axial-Flow Compressor, Turbojet Details.

Diameter & Length in inches, Weight in lb, Thrust in lbf, T/W is thrust-to-weight ratio, OPR is overall pressure ratio, Mass Flow lb/sec, SFC in lb/hour/lbf thrust. NK is not known.
* Includes afterburner.
** Avon RA 3

Data	Junkers Jumo 004	Allison J35	G.E. J47	RR Avon RA7R	RR Avon RA29	A S Sapphire ASSA7	G.E. CJ 805-3	SNECMA ATAR 9C
Diameter	32	43	36.75	42.2	41.5	37.55	31.6	39
Length	152	(195.5)*	145	102.2	124.8	125.2	188.9	(232)*
Weight	1640	2315	2554	2960	3256	3050	3213	3210
Thrust	1984	5600	5970	7500	10500	11000	11650	9440
RPM	8700	8000	7950	7800	8000	8000	NK	NK
T/W	1.2	2.4	2.33	2.5	3.22	3.6	3.6	2.95
OPR	3.14	5.1	5.35	7.1	8.4	NK	13	5.2
Mass Flow	55	91	92	130	170	155	168	150
SFC	1.39	1.1	1.014	NK	0.775	0.9	0.78	1.01
Compressor Stages	8	11	12	12	16	13	17	9
Turbine Stages	1	1	1	2	3	2	3	2
First Flight	1942	1946	1947	1949**	1956	1950	1959	1956
EIS	1944	1947	1949	1951**	1958	1956	1959	1959

Table 3

The Table below shows the comparative performances of a selection of the competing low bypass ratio engines available in the early 1960s.

Data	RR Rco 12	RR Rco 42	P&W JT3D-1	P&W JT8D-1	RR Spey Mk512
Fan Diameter	42	50	51.6	42.5	32.5
Length	132.4	154	138	123.5	109.6
Weight	4544	5000	4360	3096	2609
Thrust	17500	20370	17000	14000	12550
T/W	3.78	4.22	3.9	4.52	4.8
BPR	0.25	0.6	1.4	1.6	2.6
OPR	14.1	14.8	12.5	15.4	20.7
Mass Flow	280	367	432	315	206
SFC	0.82	0.84	0.78	0.82	0.71
Fan Stages	0	4	2	2	2
LPC Stages	7	3	6	4	5
HPC Stages	9	9	7	7	12
LPT Stages	2	2	3	3	2
HPT Stages	1	1	1	1	2
EIS	1962	1964	1961	1964	1965
Aircraft	Vickers Valiant	Vickers VC10	Boeing 707 120B	Boeing 727-100	BAC 111

In Tables 3 & 4: Diameter & Length in inches, Weight in lbs, Thrust in lbf, T/W is thrust to weight ratio, PR is overall pressure ratio, Mass Flow lbs/sec, SFC in lbs/hour/lbf thrust. NK is not known.

Table 4

The Table below shows the comparative features and performance of the initial high bypass ratio turbofans and their developed versions that made twin engine operation achievable.

Data	RR RB211 22B-02	P&W JT9-3A	GE CF6-6	RR RB211 524H	P&W 4060	GE CF6-80C2
Fan Diameter	84.8	95.6	86.4	86.3	93.6	93
Length	119.4	154.2	188	125	132	160.9
Weight	9195	8850	7896	9671	9213	9360
Thrust	42000	44250	41500	60600	60000	58950
T/W	4.56	5	5.25	6.2	6.51	6.3
BPR	4.8	5.2	5.76	4.1	4.85	5.05
OPR	24.5	21.5	25.2	34.5	31.5	30.4
Mass Flow	1385	1509	1328	1604	1705	1769
SFC	0.63	0.62	0.65	0.57	NK	0.58
Fan Stages	1	1	1	1	1	1
LPC Stages	0	3	1	0	4	4
IPC Stages	7	0	0	7	0	0
HPC Stages	6	11	16	6	11	14
LPT Stages	3	4	5	3	4	5
IPT Stages	1	0	0	1	0	0
HPT Stages	1	2	2	1	2	2
EIS	1972	1970	1971	1990	NK	1985
Aircraft	Lockheed L1011-100	Boeing 747-100	Douglas DC-10-10	Boeing 767-200 ER	Boeing 767-200 ER	Airbus A300-84-600

Bibliography and References

Images

The images used throughout this book exist in the main on the World Wide Web, and are reproduced under their original Creative Commons licence agreements, or by kind permission of the holder of the copyright of the image. The copyright applicable to this book in no way overrides the copyright of the original licence. The licence agreements for the individual images are as listed below.

Title page, File: British Airways Boeing 787-8 G-ZBJB.jpg. Source: www.flickr.com via Wikimedia Commons. Author: BriYYZ. This file is licensed under the Creative Commons Attribution- Share Alike 2.0 Generic.

Fig 1. File: Bleriot XI Thulin A 1910 a.jpg. Source and Author: J Klank via commons.wikimedia.org. Licence: This file is licensed under the Creative Commons Attribution 3.0 Unported Licence.

Fig 2. File: Anzani Military Model Fan type.JPG. Source and Author: TSRL via commons.wikimedia.org. Licence: This file is licensed under the Creative Commons Attribution-Share Alike 3.0 Unported licence.

Fig 3. File: Gnome Omega RAFM.jpg. Source and Author: Nimbus 227 via Wikimedia Commons. Licence: Released into the Public Domain via the copyright holder.

Fig 4. File: Benoist xiv.jpg. Source: commons.wikimedia.org via State Archives of Florida, Photo RC04751. The image is in the Public Domain within the US as it is time expired.

Fig 5. Roberts 6X in line aero engine reproduced from "Roberts Motor Company Aircraft Engines" by courtesy of the Author, Mr William Pearce. Source: Old Machine Press.

Fig 6. File: Vickers Vimy bomber.jpg. Source: via commons.wikimedia.org, http://www.flickr.com/photos/sdasmarchives/7304566588/in/set-72157629974581728/. Author: San Diego Air & Space Museum. This file is made available under the Creative Commons CC0 1.0 Universal Public Domain Dedication.

Fig 7. Vickers Vimy Transatlantic Route. Authors own work.

Fig 8. Rolls Royce 40/50 Motor Car Engine. (Courtesy of Autocar).

Fig 9. File: Handley Page 0 /100 aircraft.jpg. Source: Photograph from the collection of en: Library and Archives Canada, PA-125413 via commons.wikimedia.org. Licence: This image is in the Public Domain in Canada because its copyright has expired.

Fig 10. File: RRKestrelXVI.JPG. Source and Author: Nimbus 227 via Wikimedia Commons. Licence: Released into the Public Domain via the copyright holder.

Fig 11. Spitfire, File: Ray Flying Legends 2005-1.jpg.Source: fr.wikipedia; transferred to Commons by User: Padawane. Author: Brian Fury78 at fr.wikipedia. Licence: This file is licensed under the Creative Commons Attribution-Share Alike 3.0 Unported Licence.

Fig 12. File: BOAC C-4 Argonaut Heathrow 1954.jpg. Source and Author: Ruth AS via commons.wikimedia.org. Licence: This file is licensed under the Creative Commons Attribution 3.0 Unported Licence.

Fig 13. Nine Cylinder Radial Engine Configuration. This image is reproduced courtesy of Mekanizmalar. An excellent animation of the radial and rotary engines can be viewed on www.mekanizmalar.com.

Fig 14. Mustang P-51. File: P-51-361.jpg. Source: www,ww2incolor.com/gallery/albums. Author: USAAF/361st FG Association (via Al Richards) via commons.wikimedia.org. Licence: As a work of the U.S. Federal Government the image is in the Public Domain.

Fig 15. File:AU-1 Corsair in flight 1952.jpg. Source: U.S. Navy National Museum of Naval Aviation Photo No. 1986.145.002 via commons.wikimedia.org. Author: US Navy. Licence: As a work of the U.S. Federal Government the image is in the Public Domain.

Fig 16. Gnome Omega rotary aero engine installed in the Shuttleworth Collection's airworthy Blackburn Type D Monoplane. File: Gnome Omega OW.jpg. Source and Author: Nimbus 227 via commons.wikimedia.org. Licence: Released into the Public Domain by the copyright holder.

Fig 17. Cutaway Bristol Hercules engine at The Museum of Flight, East Fortune, Scotland. File: Bristol-hercules east-fortune.jpg. Source and Author: Hairyharry via commons.wikimedia.org. Licence: Released into the Public Domain by the copyright holder.

Fig 18. Wright Cyclone R-1820 at the USAF Museum; Dayton, OHIO. File: Wright R-1820 Engine.jpg. Author and Source: Highflier via commons.wikimedia.org. Licence: This file is licensed under the Creative Commons Attribution 3.0 Unported Licence.

Fig 19. Boeing B17 Flying Fortress. File: Colour Photographed B-17E in Flight.jpg. Source:www.nationalmuseum.af.mil/shared/media/photodb/photos/060515-F-1234S-018.jpg. Author: USAF Photo. Licence: As a work of the U.S. Federal Government the image is in the Public Domain.

Fig 20. File: Boeing 247D United Airlines - NC13347 - Skinnylawyer.jpg via commons.wikimedia.org. Author: In Sappo We Trust from Los Angeles, California, USA. This file is licensed under the Creative Commons Attribution- Share Alike 2.0 Generic Licence.

Fig 21. File: Pratt and Whitney Wasp.jpg. Author and Source: Sanjay Acharya via Wikimedia Commons. Licence: This file is licensed under the Creative Commons Attribution 3.0 Unported Licence.

Fig 22. File: Douglas DC-2 Uiver.jpg via commons.wikimedia.org. Author: Stahlkocher, Source: Authors Own Work. This file is licensed under the Creative Commons Attribution- Share Alike 3.0 Unported Licence.

Fig 23. File: C-47b dakota g-ampy arp.jpg via commons.wikimedia.org. Author: Adrian Pingstone (Arpingstone), Source: Authors own work. This work has been released into the Public Domain by its author, Arpingstone.

Fig 24. Pratt & Whitney R-1830 Wasp at the Imperial War Museum Duxford. File: R-1830 IWM.JPG. Source and Author Numbus 227 via commons.wikimedia.org. Licence: This file is licensed under the Creative Commons Attribution 3.0 Unported Licence.

Fig 25. Liberator Heavy Bomber in RAF Livery. File: B 24 in RAF service 23 03 05.jpg. Source: en.wikipedia. Author: Bzuk at English Wikipedia via commons.wikimedia.org. Licence: The image is now in the Public Domain because its term of copyright has expired.

Fig 26. File: Bundesarchiv Bild 101I-432-0796-07, Flugzeug Focke-Wulf Fw 200 "Condor".jpg via commons.wikimedia.org. Author: Kranz. Source: The German Federal Archive (Deutsches Bundesarchiv). This file is licensed under the Creative Commons Attribution- Share Alike 3.0 Germany. Attribution: Bundesarchiv, Bild 101I-432-0796-07 / Kranz / CC-BY-SA.

Fig 27. File: BOAC Boeing 314A Berwick landing at Lagos.jpg via commons.wikimedia.org. Author: Flt.Lt. N.S. Clark, Royal Air Force official photographer, Source: photograph CH 14069 from the collections of the Imperial War Museum. The image is in the public domain due to expiry of Crown Copyright.

Fig 28. Wright R-2600 Twin Cyclone, 14 Cylinder Engine at the USAF Museum; Dayton, OHIO. Author and Source. USAF via commons.wikimedia.org. As a work of the U.S. Federal Government the image is in the Public Domain.

Fig 29. File: Boeing 307 F-BELY Paya Lebar 1967.jpg. Source: flickr.com/photos via commons.wikimedia.org. Author: Charles M Daniels. Licence: This file is made available under the Creative Commons CC) 1.0 Universal Public Domain Dedication.

Fig 30. File: Lockheed L-1649 Constellation TWA.jpg. Source: Ames Imaging Library System (AILS) photo A83-0499-18. Author: NACA. This file is in the Public Domain because it was solely created by NASA.

Fig 31. Boeing B-29 Superfortess. File: Olive-drab painted B-29 superfortress.jpg. Source: www.af.mil//shared/media/photodb/photos/020903-0-9999b-042.jpg via commons.wikimedia.org. Author: USAF Photo. Licence: As a work of the U.S. Federal Government the image is in the Public Domain.

Fig 32. File: Wright R-3350 Cyclone Engine 1.jpg. Source: SAC Museum; Omaha, NE via commons.wikimedia.org. Author: Highflier. Licence: This file is licensed under the Creative Commons Attribution 3.0 Unported Licence.

Fig 33. File: Boeing 377 Stratocruiser, BOAC.jpg. Source www.flickr.comb & Author, San Diego Space Museum, via commons.wikimedia.org. Licence: This file is in the Public Domain.

Fig 34-.Pratt & Whitney Wasp Major at the SAC Museum; Omaha, NE File: Pratt & Whitney R-4360 Wasp Major.jpg. Source and Author: Highflier via commons.wikimedia.org. Licence: This file is licensed under the Creative Commons Attribution 3.0 Unported Licence.

Fig 35. Shut Down Rates Per 1,000 Hours, Versus Engine Horse Power, for Pre 1953 Piston Engines. Author's Own Work.

Fig 36. Whittle W1 Engine. File: Power jets W.2.jpg. Source: Flickr via commons.wikimedia.org. Author: Timitrius. Licence: This file is licensed under the Creative Commons Attribution-Share Alike Generic Licence.

Fig 37. Whittle W1 Turbojet Cross Section. Image courtesy of P P Animations via the Midland Air Museum.

Fig 38. File: IWM-CH14832A Gloster E28-39 205210674.jpg. Source: Photograph CH14832A from the collections of the Imperial War Museum via commons.wikimedia.org. Author: Devon SA, Royal Air Force official photographer. Licence: Artistic work created by the UK Government is released to the Public Domain.

Fig 39. File: Gloster Meteor Mk III ExCC.jpg. Source: Royal Air Force website/www.raf.mod.uk via commons.wikimedia.org. Author: unknown. Licence: Released into the Public Domain due to expired Crown Copyright.

Fig 40. File: Flughafen Rostock-Laage1.JPG. (Replica of Heinkel HE178 in arrival Haul of Rostock-Laage). Source: Own work by Gryffindor via commons.wikimedia.org. Licence: Public Domain.

Fig 41. Junkers Jumo 004 Axial Compressor. File: Junkers Jumo 004 Compressor Top View.jpg. Source:www.laugle.com/Jumo004.html via commons.wikipedia.org. Author: John W. Laugle. Licence: This file is made available under the Creative Commons CCO 1.0 Universal Public Domain Dedication.

Fig 42. File: Messerschmitt Me 262 Schwable.jpg. Source and Author: USAF National Museum via commons.wikimedia.org. Licence: As a work of the U.S. Federal Government the image in the Public Domain.

Fig 43. General Electric J31 Version of the Whittle W1 at the USAF Museum; Dayton, OH. File:GE J-31 Turbojet Engine.jpg. Source and Author: Highflier via commons.wikimedia.org. Licence: This file is licensed under the Creative Commons Attribution 3.0 Licence.

Fig 44. File: DH Goblin annotated colour cutaway.png. Source and Author: Ian Dunster/Stahlkocher via commons.wikimedia.org. Licence: This file is licensed under the Creative Commons Attribution 3.0 Unported Licence.

Fig 45. Lockheed P-80 Shooting Star. File: P80-1 300.jpg. Source and Author: USAF photo via commons.wikimedia.org. Licence: As a work of the U.S. Federal Government the image is in the Public Domain.

Fig 46. Rolls-Royce Nene Turbojet. File: Rolls-Royce Nene.jpg. Source: RAAF Base Pearce, Western Australia via commons.wikimedia.org. Author: Rottweiler. Licence: This file is Licensed under the Creative Commons Attribution 3.0 Unported Licence.

Fig 47, Author: Kogo via commons.wikimedia.org. Licence: GNU Free Documentation Licence (GFDL) Version 1.2, Version 1.2 or any later version published by the Free Software Foundation.

Fig 48. USSR MIG 15 Fighter at Chino, California. File: Mikoyan MiG-15, Chino, California. Source: www.flickr.com/photos/37467370@N08/7366921206. Author: Greg Goebel via commons.wikimedia.org. Licence: This file is licensed under the Creative Commons Attribution-Share Alike 2.0 Generic Licence.

Fig 49. Layout of Single Spool Axial-Flow Turbojet. File: Jet engine.svg. Source: self-made, vector version of en: Image: FAA-8083-3A Fig 14-1.PNG which comes from an FAA handbook, via commons.wikimedia.org. Author: Jeff Dahl. Licence: This file is licensed under the Creative Commons Attribution-Share Alike 4.0 International, 3.0 Unported, 2.5 Generic, 2.0 Generic and 1.0 Generic Licence.

Fig 50. Martin B-57 Canberra Bomber. File: Martin B-57A USAF 52-1418.jpg. Source: USAF Photo040315-F-9999G-006 via commons.wikimedia.org. Author: USAF. Licence: As a work of the U.S. Federal Government the image is in the Public Domain.

Fig 51. Rolls-Royce Avon RA.3 Mk.101 at RAF Museum Cosford. File: RR Avon.jpg. Source and Author: Arjun Sarup via commons.wikipedia.org. Licence: This file is licensed under the Creative Commons Attribution-Share Alike 4.0 International.

Fig 52. Wright J65, Licensed version of Armstrong Siddeley Sapphire. File: Wright J65.jpg. Source and Author: Kogo via commons.wikipedia.org. Licence: GNU Free Documentation Version 1.2 only as published by the free software foundation.

Fig 53. General Electric J79- Axial Flow 17 stage compressor. File: General Electric J79-GE-15 in Jeju Aerospace Museum 20140606 01.JPG. Source and Author Hunini via commons.wikipedia.org. Licence: This file is licensed under the Creative Commons Attribution 3.0 Unported Licence.

Fig 54. Bristol Olympus Mk 101. File: Bristol Olympus 101 gas flow diagram.jpg. Source and Author: Typandy via commons.wikipedia.org. Licence: This file is licensed under the Creative Commons Attribution 3.0 Unported Licence.

Fig 55. Model of Pratt & Whitney JT 3(J57) Twin Spool Turbojet at the National Air and Space Museum, Washington, D.C, USA. File: Pratt

& Whitney JT3.jpg. Source and Author: Sanjay Acharya via commons.wikipedia.org. Licence: This file is licensed under the Creative Commons Attribution 3.0 Unported Licence.

Fig 56, RE (SN 62-4234) in flight with full bomb load 060901-F-1234S-013.jpg. Source: USAF Museum website via commons.wikipedia.org. Author: USAF. Licence: As a work of the U.S. Federal Government the image is in the Public Domain.

Fig 57. Convair F-106 Delta Dart. File: F-106 Delta Dart 87th FIS.JPEG. Source and USAF via commons.wikipedia.org. Author: TSGT Ken Hammond. Licence: As a work of the U.S. Federal Government the image is in the Public Domain.

Fig 58. File: DeHavilland Comet.jpg. Source and Author: Adrian Pingstone (Arpingstone)via commons.wikimedia.org. Licence: Released into the Public Domain by its author, Arpingstone.

Fig 59. Aerolineas Argentinas Comet 4C over Swiss Alps, courtesy of Guy Montague – Pollock.

Fig 60. File: Boeing 707-138B Qantas Jett Clipper Johnny N707JT.jpg. Source: Boeing 707/138B/QANTAS/Jett ClipperJohnny/No7JT via commons.wikimedia.org Author: Phinalanji. Licence: Creative Commons Attribution-Share Alike 2.0 Generic Licence.

Fig 61. Rolls-Royce Conway RCo-12 Courtesy of Flight Global.

Fig 62. Pratt & Whitney JT3D-1 Modification Package. Author's own image.

Fig 63. File: 60th Air Mobility Wing, Lockheed C-5B Galaxy 87-0040.jpg. Source and Author: United States Air Force via commons.wikimedia.org. Licence: As a work of the U.S. Federal Government, the image is in the Public Domain.

Fig 64. General Electric TF39 on Lockheed C-5 Galaxy. File: ILA 2008 PD 083.JPG.Source and Author: ILA-boy via commons.wikipedia.org. Licence: The copyright holder has released this image into the Public Domain.

Fig 65. File: Aircraft engine Pratt & Whitney JT9D.jpg. Source and Author: Jaypee via commons.wikimedia.org. Licence: This file is licensed under the Creative Commons Attribution 3.0 Unported Licence.

Fig 66. General Electric CF6-6 Turbofan. File: CF6-6 engine cutaway.jpg. Source and Author: Federal Aviation Administration via commons.wikipedia.org. Licence: As a work of the U.S. Federal Government the image is in the Public Domain.

Fig 67. File: British Airways G-BNLU-2008-09-13-YVR.jpg Source, Makaristos and Author: G-BNLU-2008-09-13-YVR.jpg via commons.wikimedia.org. Licence: This image has been released into the Public Domain by its author.

Fig 68. File: Trident 62.jpg. Source and Author: TSRL via commons.wikimedia.org. Licence: This file is licensed under the Creative Commons Attribution 3.0 Unported Licence.

Fig 69. Pratt & Whitney JT8-D Cross Section Courtesy of the Transportation Board, Canada. Aviation Investigation Report A08C0108).

Fig 70. File: Delta Air Lines Boeing 727-200 N523DA.jpg. Source: en.wikipedia. Author: CFIF. Licence: released into the Public Domain by the author, CFIF at the Wikipedia project.

Fig 71. DC10-30. File: Northwest.arp.750pix.jpg. Source and Author: Adrian Pingstone via commons.wikimedia.org. Licence: This image has been released into the Public Domain by its author, Arpingstone.

Fig 72. RB211- 22B Three Shaft Turbofan. Source and Author: Rolls Royce PLC via www.rolls-royce.com. Licence: The image is copyright Rolls-Royce and is reproduced with permission of Rolls-Royce PLC.

Fig 73. Lockheed Tristar, File: Tristar-threequarters-arp.jpg. Source and Author: Adrian Pingstone via commons.wikimedia.org. Licence: This image has been released into the Public Domain by its author, Arpingstone.

Fig 74. Sud Aviation Caravelle, File: Sud SE-210.12 F-BTOE Air Inter Orly 03.08.74 edited-2.jpg. Source and Author: Ruth AS via commons.wikimedia.org. Licence: This file is licensed under the Creative Commons Attribution 3.0 Unported Licence.

Fig 75. Boeing 737-100. File: NASA TEST 737-100 prototype.jpg. Source and Author: NASA Langley Research Centre (Photo ID: EL-1996 00162) via commons.wikimedia.org. Licence: This file is in the Public Domain because it was solely created by NASA.

Fig 76. Boeing 737-800. File: Ryanair Boeing 737 (EI-ENI) departs Bristol Airport 23September2014 arp.jpg. Source and Author: Adrian Pingstone via commons.wikimedia.org. Licence: This image has been released into the Public Domain by its author, Arpingstone.

Fig 77. File: Concorde on Bristol.jpg Source and Author: Adrian Pingstone via commons.wikimedia.org. Licence: This image has been released into the Public Domain by its author, Arpingstone.

Fig 78. File: Airbus A300 cross section.jpg. Source: en.wikipedia via commons.wikimedia.org. Author: Asiir at English Wikipedia. Licence: This file is licensed under the Creative Commons Attribution- Share Alike 2.5 Generic Licence.

Fig 79. File: Lufthansa.a300b4-600r.d-aiay.arp.jpg. Source and Author: Adrian Pingstone via commons.wikimedia.org. Licence: This image has been released into the Public Domain by its author, Arpingstone.

Fig 80. File: US Airways Boeing 767-200; N255AY@ZRH;24.06.2012 657dh (7438737558).jpg. Source: Uploaded by russavia from Flickr via commons.wikimedia.org. Author: Aero Icarus from Zürich, Switzerland. Licence: This file is licensed under the Creative Commons Attribution-Share Alike 2.0 Generic Licence.

Fig 81. File: Ba b757-200 g-bpee arp.jpg. Source and Author: Adrian Pingstone via commons.wikimedia.org. Licence: This image has been released into the Public Domain by its author, Arpingstone.

Fig 82. RR 535 E4 B. Source and Author: Rolls-Royce PLC via www.rolls-royce.com. Licence: The image is copyright Rolls-Royce and is reproduced with permission of Rolls-Royce PLC.

Fig 83. Effects of 60 Minute Rule. This Image is reproduced from the ICAO Publication "EDTO Module 2" by kind permission of the ICAO.

Fig 84. Graph of Failure Probabilities. Author's own work.

Fig 85. Effect of 120-minutes rule on JFK to LHR Flight. Author's own work

Fig 86. 120 minute ETOPS Reduced exclusion zones. This Image is reproduced from the ICAO Publication 'EDTO Module 2' by kind permission of the ICAO.

Fig 87. 180 minute ETOPS, further reduction in exclusion zones. This Image is reproduced from the ICAO Publication 'EDTO Module 2' by kind permission of the ICAO.

Fig 88. Graph of Dual Engine Failure Probability Versus Diversion Time. Author's own work.

Fig 89. GE 90-115B. Image reproduced from publicly available data on www.geae.com by courtesy of GE Aviation.

Fig 90. Boeing 787 Dreamliner. File: British Airways Boeing 787-8 G-ZBJB.jpg. Source: www.flickr.com/photos/bribri/9657073746 via commons.wikipedia.org. Author: BriYYZ. Licence: This file is licensed under the Creative Commons Attribution-Share Alike 2.0 Generic Licence.

Fig 91. Airbus A350-900 XWB. File: Qatar Airways Airbus A350 on finals on its first flight to London Heathrow Airport (3).jpg. Source: www.flickr.com/photos/tagsplanepics-1hr/16061911917 via commons wikipedia.org. Author: John Taggart. Licence: This file is licensed under the Creative Commons Attribution-Share Alike 2.0 Generic Licence.

Fig 92. Rolls-Royce Trent XWB. Source and Author: Rolls- Royce PLC via www.rolls-royce.com. Licence: The image is copyright Rolls Royce and is reproduced with permission of Rolls Royce PLC.

Fig 93. Pratt & Whitney 1100G Geared Fan. Source and Author: Pratt & Whitney document Pure Power- PW1100G-JM Engine. Licence: The document is endorsed "This document has been publicly released"

Fig 94. Rolls-Royce Dart. File: Rolls-Royce dart turboprop.jpg. Source and Author: Sanjay Acharya via commons wikipedia.org. Licence: This file is licensed under the Creative Commons Attribution 3.0 Unported Licence.

Fig 95. File: Bristol Britannia RAF Museum Cosford (1).jpg Source and Author: Tony Hisgett via commons.wikimedia.org. This file is licensed under the Creative Commons Attribution-Share Alike 2.0 Generic Licence.

Fig 96. File: Bristol.proteus.arp.750pix.jpg. Source and Author: Adrian Pingstone via commons.wikimedia.org. Licence: This image has been released into the Public Domain by its author, Arpingstone.

Fig 97. Jet Engine Performance Trends with Time. Author's own work.

Fig 98. BAe 146-The Last British Airliner. File: Eurowings bae146-300 d-aqua arp.jpg. Source and Author: Adrian Pingstone (Arpingstone) via commons wikipedia.org. Licence: The copyright holder(Arpingstone) has released this image into the Public Domain.

Fig 99. File: Bell P-59B Airacomet at the National Museum of the United States Air Force.jpg. Source and Author: Federal Aviation Administration via commons.wikipedia.org. Licence: As a work of the U.S. Federal Government the image is in the Public Domain.

Books

Bleriot: Herald of an Age, B P Elliot: ISBN 0-7524-1739-8.

The Rotary Aero Engine, Andrew Nahum: ISBN 1-900747-12-X

British Piston Aero Engines and Their Aircraft, Alec Lumsden: ISBN 1-85310-296-6

The Boeing 247, The First Modern Airliner, F Robert Van De Linden: ISBN 0-295-97094-4

Lockheed Constellation: Design, Development and Service History, Peter M Bowers & Curtis K Stringfellow: ISBN 0-87938-379-8

R4360- Pratt & Whitney's Major Miracle, Graham White: ISBN 1-58007-097-3

Sir Frank Whittle - Invention of the Jet, Andrew Nahum: ISBN 1-84046-662-6

Turbojet History and Development, 1930 -1960, Anthony L King: ISBN 978-1-86126-912-6

English Electric Canberra, Bruce Barrymore Halpenny: ISBN 978-1-84415-242-1

The First Jet Airliner, The Story of the DeHavilland Comet, Timothy Walker: ISBN 1-902236-05-X

Comet-The World's First Jetliner, Graham M Simons: ISBN 975-17815-927-93

Lockheed C5 Galaxy, Chris Reed: ISBN 10-07643-120-57

The Boeing 747 Story, Peter R March: ISBN 0-75094-485-4

The Engines of Pratt & Whitney, A Technical Story, Jack Connors: ISBN 978-160086-711-8

Boeing 727, Modern Civil Aircraft, Peter Gilchrist: ISBN 978-07110-081-7

Boeing 737, Modern Civil Aircraft, Alan J Knight: ISBN 978-071101-955-3

McDonnel Douglas DC10, Modern Civil Aircraft, Alan J Wright:
ISBN 978-071101-750-6
Lockheed L1011 Tristar, Philip Birtles: ISBN 978-076030-582-9
Airbus The Complete Story, Bill Gunston: ISBN 978-076030-827-1
Boeing 757/767/777, Philip Birtles: ISBN 978-0711920-075-7
The Whispering Giant, The Story of the Bristol Britannia, Rank McKim:
ISBN 978-190223-608-7
BAE146- Modern Civil Aircraft, Michael J Hardy:
ISBN 978-071102-010-8
Beyond The Black Box, George Bibel: ISBN 978-0-8018-8631-7
Aviation Safety, Editor Hans M Soekkha: ISBN 9067642584
Turbofan and Turbojet Engines: Database Handbook, Elodie Roux:
ISBN 978-2-9529380-1-3
Theory of Aerospace Propulsion, P Sforza: ISBN 978-185617-912-6
The Jet Race of the Second World War, Stirling M Pavelec:
ISBN 978-15911-46667
Aircraft Propulsion, A Review of the Evolution of Piston Engines, C Fayette Taylor, archive.org
World Encyclopedia of Aero Engines, Bill Gunston.

Rolls-Royce Heritage Trust Publications: -
The Jet Engine: ISBN 0-902121-04-9
EAGLE-Henry Royce's First Aero Engine, Derek S Taulbut:
ISBN 978-1-872922-40-9
Rolls-Royce Piston Aero Engines, A Designer Remembers, A A Rubbra:
ISBN 1-872922-00-7
The History of the Rolls-Royce RB211 Engine, Phil Ruffles:
ISBN 978-1-872922-48-5
The Rolls-Royce Dart, R M Heathcote: ISBN 1-872922-03-1

Articles and Papers

Aviation History Online Museum Articles
Captain J Alcock and Lt Arthur Whitten Brown (aviation history.com/airmen/alcock.htm)

Boeing Model 247; Pratt & Whitney R1340 Wasp; Douglas DC-3; Pratt & Whitney R1830 Twin Wasp; Boeing 314 Clipper; Wright R-2600 Cyclone; Boeing 307 Stratoliner; Lockheed Constellation; Wright R3350; Boeing 377 Stratocruiser; Boeing B29 Superfortress; Heinkel HE 178; Junkers Jumo 004; Bell-59 Airacomet; Gloster Meteor; Messerschmitt ME 262; Lockheed P-80 Shooting Star; Mikoyen Gurevich Mig 15; Rolls-Royce Nene.

Flight Global, Flight International Articles
De Havilland Goblin November 1st 1945, Series 11 Goblin February 21st 1946, Development of the Goblin Engine May 15th 1947; Rolls Royce Nene, April 18th 1946; Canberra Bomber, December 15th 1949; The Canberra Story, 14th March 1959; The Capable Canberra, February 17th 1956; The Rolls-Royce Avon, 16th December 1955; Rolls-Royce Avon 200 Series, 11th October 1957; Armstrong Siddeley Sapphire, 6th January 1956; Armstrong Siddeley Sapphire 7, 9th November 1956; Anglo American Jewel (Wright J65 Sapphire), June 1954; Two Spool Turbo Wasp (Pratt & Whitney J57), 27th November 1953; The Comet Emerges, July 28th 1949; Comet in the Sky, 4th August 1949; The Tale of the Comet, April 25th 1952; Comet Engineering, May 3rd 1953; The Comet Accidents- History of Events 39th October 1954; Report of the Comet Engineering, 18th February 1955; Boeing 707 Revelations, 27th January 1956; Rolls-Royce Conway Bypass Turbojets, 13th January 1960; Boeing 707- Pratt & Whitney Turbofan, 1958; Pratt & Whitney JT9D Reliability, 4th May 1972; Boeing 747 Aircraft Profile; Boeing's Trimotor(727), 30th December 1960; Rolls-Royce RB211 Turbofan, 9th November 1967; Presenting the Lockheed 1011, 2nd November 1967; Flying the Caravelle, 20th December 1957; Airbus Industrie Spreads its Wings, 4th September 1975; Bristol Britannia, 12th August 1955; Bristol Proteus 18th August 1949; Power For the Giants, August 5th 1948; Aero Engines 1954, 9th April 1954; HS 146, 18th October 1973; BAE 146, 2nd May 1981; Pratt & Whitney JT3D and JT8D 23rd July 1977; VC10-What May Have Been, 18th May 1963; PIA Trident Experience, 11th May 1967; Good News for Tay, December 17th 1983;

Miscellaneous Papers and Articles

How the Constellation Became Star of the Skies, lockheedmartin.com
Pratt & Whitney R4360 Wasp Major, wikiwand.com
Frank Whittle's W2B Turbojet: United Kingdom versus United States development, enginehistory.org
Sir Frank Whittle, University of Cambridge Department of Engineering, www.g.eng/cam.ac.uk
Roberts Motor Company Aircraft Engines, www.oldmachinepress.com.
How a Radial Engine Works, mekanizmalar.com
Douglas DC-2, The Museum of Flight, museumofflight.org
Douglas DC-2, Historical Snapshot, boeing.com/history
The Jet Engine & Sir Frank Whittle, www.midlandairmuseum.co.uk
General Electric J31, National Museum of the USAF, www.nationalmuseum.af.mil
HS 121 Trident, dmflightsim.co.uk
McDonnel Douglas DC10-10, airliners.net
The Success Story of Airbus, airbus.com
Rolls-Royce Engines-RB211, gracesguide.co.uk
General Electric GE 90, geaviation.com
Trent XWB, European Safety Agency (EASA) Type Certificate TCDS EASA.E.111, easa.europa.eu
Pratt & Whitney PW1100G, Pure Power Engine Family Specs Chart, pw.utc.com
Future Civil Aero Engine Architectures & Technologies, Rolls-Royce, etc10.eu/mat/whurr.pdf
Jet Engine Specification Database, jet-engine.net
Aircraft Turbine Engine Reliability and Inspection Investigations AD-A274 860-FAA Technical Centre, dtic.mil
Behaviour of Skin Fatigue Cracks at the Corners of Windows in a Comet 1 Fuselage, Ministry of Aviation, R&M N0 3248, naca.central.cranfield.ac.uk
System Safety Assessment, Reliability of Systems and Equipment, R G W Cherry, rgwcherry.co.uk
Engine Data, Gatwick Aviation Museum, www.gatwick-aviation museum.co.uk
The Daniel and Florence Guggenheim Memorial Lecture, Civil Propulsion, The Last 50 Years (2002), ICAS 2002 Congress, icas.org/ICAS _Archive/ICAS2002/PAPERS/1.pdf

Airbus Advertising Campaign, Boeing Frontiers, Volume 1, Issue 05, September 2002, boeing.com/news/frontiers/archive

Facts About the Wright Turbo Compound, Curtis Wright Operation, Wright Aeronautical Division (October 1956), enginehistory.org/WRIGHT/TC%20facts.pdf

Statistical Summary of Commercial Jet Airplane Accidents, boeing.com/resources

Regulations and Procedures

ICAO Uniting Aviation, Module 2 Basic Concepts; Module 4 Aircraft Certification Considerations, www.icao.int

Extended Operations (ETOPS) and Polar Regions FAA, AC-120-42B, faa.gov/regulations

Extended Operations (ETOPS), Eligibility for Turbine Engines, Advisory Circular AC No 33.201, faa.gov/documents

FAA New ETOPS Regulations, faa.gov

Federal Register-Extended Operations (ETOPS) and Multi Engine Airplanes, federalregister.gov

ETOPS, Extended Range Operations (ETOPS) With Two Engine Airplanes, globalsecurity.org

The New ETOPS Rule, Boeing, boeing.com

EASA (European Safety Agency), Extended Range Operations with Twin Engine Airplanes, AMC20-6 rev2, easa.europa.eu

FAA Advisory Circular AC 25.1309-1, System Design Analyses (also JAR 25-1309), www.faa.gov/regulations

Air Carrier Certification FAA 14CFR121

Aviation Rule Making Advisory Committee (ARAC), Extending Operations Working Group Recommendations, ACO ETOPST1-061400.pdf, www.faa.regulations_policies

ICAO Report, AN-WP/5598, Extended Range Operations of Twin Engine Commercial Air Transport Aeroplanes, 15 Feb 1984.

AD-A274-8